The Global Citizen

The Global Citizen

A Guide to Creating an International Life and Career

Elizabeth Kruempelmann

Ten Speed Press
Berkeley/Toronto

Ten Speed Press
Box 7123
Berkeley, California 94707
www.tenspeed.com

Distributed in Australia by Simon & Schuster Australia, in Canada by Ten Speed Press Canada, in New Zealand by Southern Publishers Group, in South Africa by Real Books, in Southeast Asia by Berkeley Books, and in the United Kingdom and Europe by Airlift Book Company.

Cover illustration and design by Kathy Warinner
Text design by Lynn Bell, Monroe Street Studios

Library of Congress Cataloging-in-Publication Data

Kruempelmann, Elizabeth.
The global citizen : a guide to creating an international life and career / by Elizabeth Kruempelmann.
p. cm.
ISBN 1-58008-352-8
1. Travel. 2. Employment in foreign countries. 3. Foreign study. I. Title.
G151 .K63 2002
650.14—dc21
2002004191

Printed in Canada
First printing, 2002

1 2 3 4 5 6 7 8 9 10—06 05 04 03 02

Dedication

To Frank, Nicholas, and Baby #2 with love

CONTENTS

Acknowledgments

My global pursuits, and this book, would have never become a reality if it weren't for the encouragement, help, and support of many family members, friends, and colleagues. First, a special thanks to my generous parents, whose love and personal sacrifices have allowed me to live out my dreams of seeing the world. I'm also particularly grateful to my wonderfully devoted husband, Frank, for giving me the opportunity to live the dream life abroad I've always wanted, and whose constant love and support has instilled in me the courage to make this book a reality.

My sincere gratitude to the many people who have helped me with this book:

Eve Berton, who, in the fall of 1998, inspired me (over dinner at an Israeli restaurant in Warsaw) to write a book. As a writer and traveler herself, she encouraged me to capture my adventures and share them with the world. In this book, she has generously shared her experiences as an artist/writer living and working in Poland.

Michael Landes, author of the awesome *The Back Door Guide to Short-Term Job Adventures* (Ten Speed Press, 2000), who asked me to write an article for his book in 1999. Ever since then, he has been my authoring mentor and friend, and although I have only met him once in person, he continues to generously guide and encourage me through our regular emails.

Sarah Seeland, who became a dear friend in 1992, during our initial adventures in Germany, and who has contributed articles to the Working Abroad chapter about what it takes to succeed at finding a job abroad.

Chrissy Picardi, with whom I shared many unforgettable college moments and subsequent international "Thelma and Louise" travel adventures, and who has contributed her Semester-at-Sea and language-learning adventures to this book.

Michele Carlson, a model global citizen if I ever knew one, and with whom I bonded during my first experience studying abroad in Copenhagen in 1989–90. She has contributed an article about her Peace Corps and teaching experiences.

Claire Campbell, a brilliant friend of a friend who has generously shared her study-abroad adventures.

Rebecca Falkoff, a former Monster.com colleague and fellow writer who shares her insightful experiences of living abroad.

Mark Thompson, a brilliant and very funny ex-lawyer who shares his story of rediscovering his global interests.

Bridget Johns, whom I met during our work adventures in Poland, and who has graciously written about her life working abroad.

Laurel Eismann, a fellow-American friend living in Portugal who shares a story about how she ended up working overseas.

Jean Blomquist, my copyeditor, whose keen attention to detail has put the finishing touches on this book.

Lynn Bell, who creatively pulled together the various elements of this book into an attractive design.

Kathy Warinner, a great cover designer.

Julie Bennett, my editor at Ten Speed Press, whose invaluable comments, guidance, and direction have made this book stronger and better than I could have imagined.

Ten Speed's publisher, Kirsty Melville; editorial director, Lorena Jones; and sales manager, Gonzalo Ferreyra; for getting this dream of mine out into the world!

Preface

The traditional rules of living and working have changed for good. The world is becoming more flexible to fit our more demanding lifestyles. We love the Internet and all those cool techno-play gadgets so much that we can't imagine how we ever made do without them. As we anxiously await the next exciting lifestyle leap that all the latest and greatest technological wonders will bring about, you might be wondering about leaps of a different nature—leaps involving airplanes and passports. "What about a life where I'm living, working, and playing all over the world? You know . . . telecommuting from Bali, retiring in Italy, and traveling everywhere else in between?"

Despite the innovative technology that makes life easier and more mobile, for many of you the missing piece of a "perfect" life puzzle is living, working, and traveling abroad. The mystery and excitement of what lies beyond the familiar borders has captivated our imagination since the beginning of time—this phenomenon is nothing new. What is new, however, is that it is finally *your* turn in the cosmic cycle of things to strike out and discover countries, people, languages, ways of doing business, and many other aspects of life that are the current state of affairs in the world. You have already dreamed about your future; now is the time to gather your creative juices and make it happen.

With a bit of planning and some careful research, you can set the compass in the right direction: toward that special place on the globe where you will begin, or maybe continue, your global adventures. And that's where I come in. When I sat down to write this book I wanted (it) to accomplish two things. The first was to inspire you to embrace the world by showing you how to get international experience, and the second was to help you take your international experience to the limit, creating an exciting international life for yourself.

Twelve years ago, before I ever set foot on foreign soil, I couldn't have imagined that someday I would live and work in Denmark, Germany, Portugal, and Poland, converse in three languages, travel to exotic places like Moorea, Zimbabwe, and Nepal, and have a cool job that I can do from anywhere in the world. But, if I can do it, anybody can do it—and that includes you!

So, what prompted me to write this book? Well, it's quite simple really. Since studying abroad in Denmark and Germany, I've had an insatiable craving to live and work in foreign countries. Why? Because it is fun and stimulating, and I feel more alive and challenged when I'm in exotic surroundings. But to make this dream a reality, I needed guidance. I searched every bookstore in town for books explaining how to live, work, and build a life overseas. Although there were a few books on how to find a job abroad, they merely described different industries where international jobs could be found. I was looking for much, much more. In the end I discovered that no book actually offered realistic and clear guidance to getting my eager, adventurous soul out into the great big world.

Since I'm not the type of person who lets a little lack of knowledge or comfort prevent me from getting out there and just doing it, I just did it—and figured out the rest along the way. That's how I ended up back in Germany, and what happened after that was nothing less than a dream come true. I met someone (who is now my husband) with whom I could share my passion for the world and whose career has enabled us to take advantage of many global opportunities. We visited quaint European towns and medieval villages, rode camels in the Sahara Desert, went scuba diving in the Caribbean and snorkeling on a remote island in the South Pacific, took a wildlife safari in Zimbabwe, trekked through the Himalayas—and that was just the travel part! Today we truly lead a global life, living and working in various countries—two expats with a new baby on the way and a toddler in tow.

OK, so I can't guarantee that you'll meet the perfect mate, get married, and live happily internationally ever after. But it happened for me, and with a little persistence and an open mind, who knows, it might happen for you too. I have to admit, my "fairy-tale" life didn't just happen overnight. As I fumbled through each new country, figuring out how to create this global life for myself, I continued to search high and low for a guidebook, reading everything from career guides and work- and travel-abroad books to self-development literature. Still, there was no one practical guidebook that understood the reality I was experiencing living overseas, or that spoke to the needs of those of us who wanted to build an international life and global career.

Well, you know the saying, "If you really want something done, do it yourself!" And so, once again, I just did it. After twelve years and four foreign countries, discovering the ins and outs of traveling, studying, and working overseas, I figured it was about time for me to share my hard-earned wisdom, practical tips, and extensive resources with other adventurous souls, like you, who are out to discover the world. The information contained within these pages comes from five sources: my own adventures abroad, travel- and work-abroad books, the latest Internet Web sites dealing with living abroad, unique stories from friends who have lived overseas, and my experience advising people who are pursuing a global life.

As the International Career Mentor coaching job seekers on Monster.com's international job board, I became acutely aware of the real needs of people who are passionately, sometimes desperately, pursuing international experience and a global life—after all, I was in their shoes not so long ago. By helping you "internationalize" your career planning and life management, I want to make it easier for you to live out your global dreams. My hope is that *The Global Citizen* will not only answer your questions, but also help you live as a true global citizen.

How to Use This Book

In *The Global Citizen: A Guide to Creating an International Life and Career,* you will discover that the global citizen's path is a viable, and increasingly popular, life choice for people of all ages. *The Global Citizen* is more than just another resource for traveling, studying, working, and volunteering abroad, although it offers the best resources for these as well. This book is unique in that it expands on the new emphasis many people are putting on the integration of international experience into their lifelong personal and professional goals. It focuses on a common desire to experience the world in a meaningful way through educational travel, academic learning, volunteering, or professional experience overseas. Not only is international exposure desired by students, professionals, families, and retirees, but it is also establishing itself as the new threshold for career success and lifelong development. The practical experience you gain by living overseas makes this book a must for anyone interested in living life to its fullest.

There are international experiences available for almost every personality, hobby, profession, time frame, and budget. This book is composed of organizations and contacts that support the global citizen philosophy, offering the best international resources to help you find a fulfilling overseas adventure.

The Global Citizen's life-planning approach has been developed to lead you every step of the way to designing your own unique global existence. It is divided into three parts, each containing guidelines and resources to assist you in planning your overseas experience and in applying your newfound skills and wisdom toward living a more rewarding life.

For those of you who like to plan according to your goals, the best way to use this book is to read it straight through. In Part I—How in the World Do I Get Started?, you will first evaluate your motivations for going overseas. This is where you set your compass in the right direction, taking an introspective look at yourself and your goals. A unique feature of this section is the Culture Prep, a cross-cultural mini-course and self-assessment that will help you identify your cultural type, offer insights to cultural classifications, and even prepare you for culture shock (see chapter two).

The options for gaining international experience are presented in Part II—A World of Opportunity. Beginning with an overview of all of the exciting ways you can travel and live abroad—educational travel, learning overseas, volunteering, and working abroad—you will review all of your options in each category, learn about the personal and professional benefits that each experience offers, and become aware of the potential pitfalls. Each chapter includes detailed advice, the best Internet links, and other resources to help you choose the option that will meet your current and future goals.

As you read through Parts I and II, you will probably find several different options you'd like to pursue, like an adventure tour in Peru, a study program in China, or a work opportunity in Germany. Then, in Part III—Making Your Global Dream a Reality: Practical Stuff You'll Need to Know, you'll get down to the nitty-gritty details about things you should know as you embark on your journey. Chapter nine will show you how to make your dreams of living, traveling, and working abroad happen financially. Chapter ten gives you a list of predeparture tasks, chapter eleven tells you how to maximize your time abroad, and chapter twelve explains the kinds of situations you might encounter when you return home.

The last chapter of the book shares ways to contribute as a global citizen. Refer back to this section after you've ventured forth into the world. By that time you will have manifested your global passions and joined the worldwide community of global citizens. In the course of your adventures, I hope you will share your worldly knowledge and contribute in your own unique way to making the world a better place. Contributing to the world as a global citizen is what it's all about!

Boxed feature articles, called "A Global Citizen's Perspective," are success stories from ten global citizens who will show you realistically what to expect in your gallant pursuit of international experience. Other boxed articles and sidebars provide additional information on certain programs or topics of particular importance.

There is an overwhelming amount of Web sites, programs, books, and other resources that cover international topics such as traveling, studying, working, volunteering, and even retiring overseas. I don't intend to simply reiterate the information that is already out there. Rather, my intention is to make you aware of the characteristics most successful global citizens have, show you how to gain international experience, and help you live a more successful and fulfilling life yourself.

Keeping in mind the financial considerations of my readers, I put enough solid information in this book so it will not be necessary to invest money in other books or resources if you have a reliable computer and an Internet connection. In addition to following the guidelines in this book, you can usually find any other information you need for free on the Internet. (Other books on studying, working, or volunteering abroad generally consist of an overwhelming list of program names, which can now be found more effectively, efficiently, and for free online.) *The Global Citizen* Web site (www.the-global-citizen.com) is meant to be a supplement to the international life- and career-planning information in this book. The site includes links to self-assessments, programs for living abroad, and an international community of global citizens.

The Internet can and should be used for planning your overseas adventure for several reasons. First, you will have instant access to information that is more comprehensive than any book can provide. Second, by searching databases you can narrow your research quickly and more precisely than you could otherwise. Third, on the Internet you can send away for more details or program applications, apply for international jobs, get travel information, book flights, thoroughly research your destination, and many other things that will make planning easier. All of the Internet sites listed in this book are leading sites in their categories and have been tested for reliability. If you do not have Internet access, I have provided the names of books and reference materials that can be found in most libraries and have listed contact numbers you can call to receive free information in the mail. Also, many public libraries offer free Internet access, so try logging on there.

We are all citizens of this world. It is our home, and we have the privilege and the responsibility to make the most of it while also contributing to its betterment. *The Global Citizen* will help you take advantage of opportunities that will affect your education, career, hobbies and interests, personal relationships, and active causes. As this book guides you on your journey to discovering a world of opportunity, I hope you will share the information in the following pages with others. Together we'll learn to live not only as citizens of our home countries but also as true citizens of the world.

INTRODUCTION

Anyone with a passionate desire to enrich their life through a variety of international experiences can become a global citizen. Throughout my life abroad I've met inspiring individuals and couples who have chosen to travel, live, or work overseas because they believe life is too short not to take advantage of what the world has to offer. Their journeys have transformed them into truly global-minded people. If you want to live a more successful and rewarding existence (and I assume you do or you wouldn't be reading this book), then get ready for an exciting, challenging, and life-changing journey that will transform you into a global citizen too.

So, What Is a Global Citizen Anyway?

Global citizens are global-minded people like you and me who crave international experience and are passionate about living fulfilling lives. The term "global citizen" creates awareness of a whole category of internationally oriented people who derive satisfaction from life by discovering the world. Global citizens can come from any part of the globe—Canada, Europe, China, Ecuador, Singapore—you name it. By living in foreign countries, global citizens tend to form a unique cross-cultural group. Their worldly outlook on life bonds them together with like-minded thinkers who appreciate the world at large—its people, cultures, history, engineering marvels, natural resources, and all the fascinating facets of life that make the world an exciting place. The global citizen's philosophy is based on the awareness that stimulating experiences of living in foreign countries help us develop as people. As we clarify our understanding of ourselves and our world, we improve the quality of our lives.

Life sometimes passes by quicker than we realize. We study, work, travel, raise families, maybe volunteer somewhere, hopefully enjoy life, then die. If you feel a slight twinge of envy when you hear about a friend's overseas adventures or watch a travel documentary on television, it is because you know how fulfilling an experience in a foreign country can be. I hope to make your life journey more enjoyable by increasing your awareness that you can choose to live as a citizen of your own country as well as a citizen of the world.

In our home countries we are generally encouraged to get an education, work, volunteer, and travel. As citizens with the right to travel to almost any country, we have the opportunity, and I dare say, the obligation, to study, work, volunteer, and live in other parts of the globe. Yet relatively few of us know much about the world except for the country in which we live.

You might be asking yourself, "So, what types of people go abroad and how do they manage to lead these seemingly exotic lifestyles?" There is really nothing magical about it. In fact, living a global life is simply a matter of choice. Becoming a global citizen means experiencing life beyond your country's borders. The first step involves taking initiative to choose an education, a career, and hobbies that will prepare you for a dynamic, global lifestyle. By doing so, you can start to acquire the skills and characteristics that are common to people who have been successful in carving out international paths.

The Characteristics of a Global Citizen

I have observed a consistent set of traits in people who seem to live life to its fullest. These global citizens have prepared themselves with the right skills to be able to live, work, and get along well almost anywhere. What makes many of these people particularly unique is that they also take personal responsibility for bettering the world and helping the people in it.

Unless you were raised with globe-trotting expatriate, military, or missionary parents, developing the characteristics of a global citizen may take some effort. But don't let that scare you! The time and effort you invest in creating your own unique international experiences, where you develop a new perspective on the world and savvy global skills, will create the foundation of your future international path. The good news is that you probably already possess some of these traits. They can be little things, like keeping up on world events by watching CNN, or being able to remember your college French well enough to feel confident chatting with any Frenchman. Maybe you even have previous experience studying, working, or traveling in Europe, Asia, or South America. All these things will come in handy as you evaluate your status as a global citizen.

For the characteristics you would still like to acquire or strengthen, nothing beats the thrill and hands-on experience of developing these traits abroad. Acquiring the following set of skills and traits will put you on the path to international life and career success.

- Curiosity about the World
  Would you love to know what it's like to live in Spain or Italy? Are you lured by the fun-loving people, delicious food, and laid-back lifestyle? People who want to live an international life need to be curious about what lies beyond their borders: the people, cultures, mentalities, history, geography, and natural wonders. A strong desire to discover what the world has to offer will enhance your international adventures. Sometimes you need to see what you've been missing by going abroad first, but that trip inevitably spurs an intense curiosity to learn more.

- Hobbies and Interests with an International Element
  Developing hobbies and interests with an international element further adds to your international persona. Latin dancing, cooking ethnic foods, following international sports teams—hobbies with a global aspect give you a point of reference for creating a life that incorporates your interests. Chances are if you like Latin dancing, you will be able to appreciate and adapt to living or working in Latin cultures. And if you are a soccer fan, you'll have no problem bonding with sports lovers in the rest of the world.

- Awareness of Global Issues
  As a global citizen you want to be aware of the issues affecting you, your country, and other countries around the world. Staying current is easy if you read an international newspaper like the *Financial Times* or watch an international news channel like *CNN International*. Global issues can range from world finance and environmental concerns to regional issues like Middle Eastern politics and the AIDS epidemic in Africa. Defining the issues that interest you the most will help you target future involvement with these issues and influence where you decide to travel, live, or work.

- An International Network of Contacts
  They always say it's not what you know, but who you know, and that's as true around the world as it is in the States. International contacts are essential—both as personal friendships and for professional connections. Global citizens tend to have a solid network of people with international experience that they can rely on for business connections and career advice. You can start building up your network in your home country. The key is to keep this network alive throughout your life because you never know when you will need to draw from it.

- Involvement in and Contribution to Global Causes
  Many citizens of the world are aware of global problems, from preserving rain forests to feeding starving children, and they do something about it. Although you might not be involved with any global causes right now, it's likely that while you are out in the world gaining international experience you'll have a humble moment when you are exposed to one of the world's injustices and decide it is time to take a stand. Your contribution can be simple, like donating money to UNICEF or, on a larger scale, like participating in a climbing expedition of Kilimanjaro to raise money for CARE (see page 358). The important thing is that at some point you take responsibility to improve the world for future generations. As you will discover in Part III of this book, there are many ways for you to get involved with your favorite causes.

- Language Proficiency
  Learning a foreign language, or several languages, if you are linguistically talented, aids you significantly in becoming bi- or multicultural as well as bi- or multilingual. Opportunities to study a foreign language abroad are available to almost everyone. Official language tests to measure ability in certain areas like commercial terms or technical knowledge are given by several foreign organizations (see Chapter Five: Learning Abroad). Language proficiency and the ability to function across cultures affect the amount of respect you receive from people in other countries and make for a more fulfilling experience overall.

- Cross-Cultural Appreciation and Adaptation
  This type of appreciation can only be learned by directly experiencing a foreign culture. As you adapt to the differences of living in a foreign country, you learn to value other people, mentalities, and cultures. Language fluency will help you gain cross-cultural appreciation, and the ability to adapt to local ways of behaving will help you truly appreciate each culture for what it is.

- A Global Way of Thinking
  Thinking globally means you are able to analyze situations and problems from the point of view of other cultures. From a business standpoint, that means you think in terms of global opportunities for your company and career. It can also mean that you are keenly aware of the issues the world faces, such as prejudice, environmental concerns, and social injustice. Living in a foreign country for a period of time will help you understand how many issues, from religion to politics, can be seen from different, yet equally valid, points of view.

- International Travel Experience
  Global citizens tend to engage in travel experiences that are educational as well as relaxing and pleasurable. Hence, travel is just one more way to add to your knowledge about the world. Most people who have lived abroad also traveled in the country or region they were living in, thus adding to their cultural perspective of that area. (Learn more about educational travel in chapter four.)

- An International Education
  This implies taking a course, studying abroad, or participating in a fellowship program, a professional exchange, or specialty skills training. Interaction with teachers and students from different countries stimulates learning and offers insights from sometimes dramatically different points of view. Many of the most successful people who live and work abroad have also studied for a time overseas. An international education is an excellent way to get started with a global career. (Learn more about educational opportunities overseas in chapter five.)

- International Work Experience
  Whether you are volunteering or gaining professional work experience in the form of an internship, a full-time work assignment, or a long-term career, working abroad can enhance your perceptions of the world. Many people seek out shorter-term international travel, education, volunteer, or work experience to prepare for a global career that allows them to spend multiple years working abroad. (Read more about volunteering and working overseas in chapters six through eight.)

CHALLENGE YOURSELF TO BECOME FLUENT IN A FOREIGN LANGUAGE

Sprechen Sie Deutsch? Fala Português? ¿Habla Usted Español? In most developed (and many lesser developed) countries of the world, students are taught to read, write, and speak at least one foreign language proficiently. In Scandinavia, it is common for students to speak up to six languages conversationally or fluently. Most northern Europeans have an impressive knowledge of English as well as a third, or possibly fourth, foreign language. I was amazed to learn this when I first went abroad to study in Denmark.

My fourteen-year-old Danish neighbor, Lars, could already easily converse in Danish, English, Swedish, Norwegian, and German, but claimed his French wasn't that good yet. I was convinced this kid was a child prodigy until I met more Danish people. In their modest Danish way, they confessed that multilingual fluency is the norm in Denmark, and Lars, in fact, had no special language abilities that other Danish kids did not also have. Apparently multilingualism is important for a small country like Denmark, which has needed various languages to survive economically throughout the centuries. The fact that Danish children begin learning foreign languages as soon as they start school has often made me wonder why American children aren't also made to learn foreign languages at an early age when they are most likely to learn quickly. After seeing for myself the great advantages that foreign language fluency offers, both personally and professionally, I decided it is worth going into this subject in a bit more detail.

- Sharing International Experiences
  Some travelers like to give back to the world by joining organizations with a certain mission, volunteering for a world cause, or becoming an activist for their favorite organization. You can write an article for a local newspaper or talk with high school students about the rewards of gaining international experience. Your contribution can also be very personal, such as sharing your experiences and photos with family and friends. The main point is that you share your experiences in a way that will positively influence others.

In most countries of the world, like in Denmark, learning a second or third language is not an option, but rather is taken for granted as a necessary tool in life. However, as I found out after living in Europe for a few years, native English speakers who are not proficient in a foreign language have a distinct advantage in some respects because English is the language of international business. On the other hand, limiting yourself to knowing only your native language limits your interpretation of the world and your access to a wealth of interesting opportunities, both in your home country and abroad.

Even though English is spoken by many people worldwide, you should still try to become fluent in at least one foreign language. Learning a language can be a very humbling experience, especially for adults. It's even more challenging in countries like Denmark and Germany, where many of the locals prefer to speak to you in English unless you speak their language flawlessly. Language fluency is a necessary step to achieving a degree of international savvy and to being on par with your international counterparts. Plus, there is something oddly satisfying about being able to speak the local language better than the locals speak English.

Of all the language facilities, speaking is the most important and salable skill, although writing and reading can be quite important as well, depending on your profession and interests. The only way to truly learn a language is to speak it at every opportunity. If people speak back to you in English, which happened to me frequently in Denmark and Germany, continue to speak to them in their own language. They are probably trying to practice their English as much as you are trying to practice their language! It might be an uncomfortable battle of wills at first, but I have found that sticking to your guns will win respect from the locals in the long run. Despite the initial discomfort, avoid speaking English and try to speak only in the local language. Eventually, using the foreign language will become a habit (even if all you're saying is "I don't understand. Would you mind speaking more slowly?"). This type of cultural immersion is integral to understanding not only the literal language but also the cultural meaning behind it.

Foreign language ability also enables you to have a more enriching personal experience traveling in the countries where that language is spoken. For example, if you know Spanish, you can talk your way around Spain, and most of Central and South America. Language ability can also open doors for you in the professional arena. In addition to being able to converse with the neighborhood shopkeepers, read the local newspapers, and understand the morning radio announcer, think about how many more job opportunities or how much more earning potential you might have if you know one or more foreign languages. Foreign language fluency is the entrance to whole new worlds of understanding—it is up to you to open the door and go inside.

If I haven't convinced you yet, here are ten benefits of learning a foreign language.

Ten Benefits of Learning a Foreign Language

1. Understanding the language puts you in a better position to understand a culture's mentality, history, and traditions. You will be able to converse with the locals, understand television and radio, and read local papers, books, and literature.

2. Language ability shows that you are making an effort to educate yourself about the environment in which you are living. It earns you respect, giving you closer access to the people and culture.

3. Learning your first foreign language will make it easier for you to learn more foreign languages. If your first foreign language is a Romance language like Spanish, you will already have a head start on learning other Romance languages like Italian and French; the same is true for Germanic, Slavic, and other language groups. Even if the languages you learn are from different language categories, your basic understanding of how to learn a foreign language will aid in your language assimilation.

4. Believe it or not, learning a foreign language will give you a new frame of reference for understanding yourself and your own language, culture, and traditions. Seeing your situation, behavior, and norms through the eyes of another culture can be rather insightful.

5. Knowing a language (like Spanish, French, Arabic, or German) that is spoken in several countries, gives you that much more mileage for your effort. You will be able to expand your cultural understanding of several countries as they relate to each other by paying attention to how the language is adapted to reflect each country's individual character.

6. Foreign languages are your keys to a variety of social, educational, and business opportunities that you may not otherwise have access to. If you want to study or do academic research through an all-expenses-paid Fulbright or Rotary club scholarship, foreign language ability will be a plus in getting your application considered.

7. Having direct experience with a foreign country and its language solidifies the connection you'll make between that region of the world and current or historical events or personalities. I pay much closer attention to news events that happen in places I've visited.

8. Languages are tools for becoming more adaptable to an environment. They allow you to alter your perceptions and achieve greater understanding. This means letting go of your ethnocentric ideas and adopting a more open view of the world.

9. Language proficiency builds confidence. My knowledge of Danish prepared me for learning German. The similarities in the languages gave me a bit of a jump start on German and accelerated my confidence to buy train tickets, ask for directions, order from a menu, and eventually converse on a variety of topics.

10. Foreign language ability puts your knowledge, skills, language, and culture into perspective. You'll be able to understand your country's issues from an international viewpoint and get a better grasp on global issues, from world poverty to international politics.

To give yourself the best chance of success on the international scene, focus on obtaining the global characteristics you lack. For example, if you're striving for a global career, you maximize your chances for success by having an international degree, work experience in your field overseas, foreign language skills, and international contacts. On the other hand, if you are looking for a global adventure to fulfill a personal hobby in archaeology, for example, you might want to combine an educational travel expedition to Egypt with volunteering on an archaeological dig, building friends and contacts for the future.

Why Become a Global Citizen?

I am convinced that all people, regardless of age, personality, or nationality, benefit from pursuing interests outside the confines of their home countries. International exposure helps you plan your life by opening up a whole new set of options. If you make the effort to get international experience you are virtually guaranteed to get something out of it that will improve your life (more money, marketable skills, enjoyment, knowledge, degrees, an opportunity to make a difference), enhance your career success, generate adventure, give others hope, or provide you with new knowledge, skills, and experience. Of course, this means leaving your comfort zone for the unknown—and that means new wisdom, personal growth, and a very fulfilling life.

Fitting an International Experience into a Crowded Life

Whether you are planning for yourself, your children, your parents, or a physically challenged person, there are many ways you can start integrating international experience into your lifelong adventures. Of course, more opportunities will be available the earlier you start developing international skills. Teenagers and high school students can begin to increase their global awareness through cultural homestays, summer jobs in a foreign country, an educational trip organized for students, or a study-abroad program. College students and graduates can pursue these same types of international exchanges as well as arrange overseas internships and professional training programs, or by volunteering or taking a course of study overseas.

With the plethora of international opportunities available today, people of all ages can pursue experiences that used to be reserved for adventurous students and young people. For example, retirees can take a variety of courses at international universities, and families can work together on community service projects in developing countries. Professionals from all fields can teach English or participate in professional exchanges abroad, and job changers can use the time between jobs to volunteer or take part in global seminars and educational trips.

International experience can be pursued at any age and in many different ways. Although we'll explore all of our options in detail in Part II, here is a brief sampling of how some of the programs can be enjoyed at each stage of life.

International Opportunities for All Ages

High School Students

Life Stage	International Opportunities Available	Benefits of International Experience
Tina, 18-year-old high school senior Tina likes languages, but is not sure what she wants to study in college. Through a Rotary Club scholarship, she takes a year off before college to live with a family in Germany. Tina attends a local college to study the German language and works part-time in a local German pub. After the experience, she decides on a career direction. Her language fluency and international experience help her get a scholarship to a top college, where she studies medicine. Tina eventually goes abroad again to work in an international health clinic.	Rotary Club Scholarship to study abroad (see page 308) AFS Student Exchange (see page 138) CHI Cultural Homestay (see page 98) Berlitz Language Vacation (see page 102) SCI-IVS Volunteer Work Camp (see page 196)	Develop independence Grow emotionally and mentally Learn problem-solving skills Gain language skills Receive preference with college admissions and qualify for scholarship opportunities

College/University Students

Life Stage	International Opportunities Available	Benefits of International Experience
Lisa, 21-year-old college junior Lisa is a business major and Spanish minor who wants to do an internship abroad. She arranges a summer internship in Mexico. After improving her Spanish and narrowing down her business interests, Lisa is well prepared to do further internships abroad, giving her a jump start on a global management career.	AIESEC Professional Internship (see page 233) CDS Traineeship (see page 234) CIEE student travel and language course (see page 127) Solo adventure travel (see page 108) WorldTeach (see page 212)	Increase awareness of what lies beyond one's borders Develop cross-cultural understanding Gain marketable skills Practice language skills Solidify professional direction Increase chances of graduate school admissions, scholarships, and job offers

Graduate and Postgraduate Students

Life Stage	International Opportunities Available	Benefits of International Experience
Mike, 24-year-old graduate student Mike wants international work and language experience. He joins the Peace Corps and does volunteer work in Africa for two years. Mike learns Swahili and gains teaching and project management experience. He uses his network of Peace Corps and international contacts to develop an international career with a nonprofit organization.	Peace Corps (see page 209) Global Services Corps (see page 200) MBA Management Trainee Programs (see page 152) Graduate Degree from School for International Training (see page 147) Global Exchange/Reality Tour (see page 107)	Same benefits as college students, plus Get relevant professional experience Start building a global network Do something different before entering the world of full-time work

Job Changers

Life Stage	International Opportunities Available	Benefits of International Experience
John, 30-year-old single engineer John is interested in archaeology. He participates in a two-week, summer volunteer archaeological dig in Egypt. He gets hands-on experience, meets new friends, and makes contacts in the archaeology field. John continues to find new ways to explore his hobby through travel, courses, and volunteer work. He eventually gets a new job combining his engineering skills and his archaeological interests.	Earthwatch Volunteer (see page 199) Robert Bosch Fellowship in Germany (see page 162) Teaching Overseas (see page 239) MIM Degree at Thunderbird (see page 155) Create a Job Abroad (see page 280)	Same benefits as students, plus Consider a different professional direction Establish a network of international contacts Be challenged by new adventures Experience renewed personal and professional fulfillment

Working Professionals

Life Stage	International Opportunities Available	Benefits of International Experience
Martina, 35-year-old physical therapist Martina is devoted to the health-care profession. With her boss's approval of a one-month leave, she joins health volunteers in Africa. Gaining awareness of the world's health problems, she adds depth to her skills, is challenged by third-world working conditions, and solidifies her devotion to her profession. Eventually Martina donates more time to helping local and international health-care organizations.	Peacework (see page 203) Get a Job Overseas (see page 217) Career Development Program through AIPT (see page 233) Professional Exchange through CIF (see page 166) Rotary Grants for University Teachers (see page 162)	Advance professionally, earn more money, collect new knowledge Learn international skills Make industry contacts worldwide Increase cross-cultural awareness Increase understanding of how your profession is practiced in other countries

Families

Life Stage	International Opportunities Available	Benefits of International Experience
The Smiths, married couple in their mid-40s with teenage children and a tight budget Mrs. Smith organizes group travel with another couple and earns free travel for herself. She also saves frequent-flier miles and finds special travel deals on the Internet. The Smiths chaperone six teenagers on a twenty-day tour of Western and Eastern Europe. The teenagers and adults gain an awareness of foreign languages, cultures, and world issues. The Smiths are happy they found a way to expose their children and themselves to the world.	AIFS Group Leader (see page 320) Frequent-Flier Miles (see page 322) Journeys International Family Trips (see page 96) SERVAS Homestay (see page 99) Community Volunteering through Global Citizen's Network (see page 196)	Bond as a family Have fun in a unique and unforgettable way Learn about other cultures Visit cultural sights Practice a foreign language Teach kids about the value of money

50+ Crowd and Retirees

Life Stage	International Opportunities Available	Benefits of International Experience
The Millers, retired couple in their early 60s Recently retired, the Millers now have time and money to travel abroad for the first time. They join Elderhostel on an educational religious tour in the Middle East. They meet other senior couples and feel motivated to have a fulfilling retirement. The Millers get involved with other special tours and volunteer programs abroad for retirees.	Eldertreks Educational Travel (see page 113) Elderhostel Courses (see page 113) Walking the World (see page 114) International Executive Service Corps (see page 189) Semester at Sea (see page 148)	Stay physically and mentally active Meet people from around the world Have fun Be able to give back to society by volunteering

Disabled Travelers

Life Stage	International Opportunities Available	Benefits of International Experience
Bob, 43-year-old physically challenged man Bob is enthusiastic about the prospect of teaching abroad through MIUSA. He is accepted as a Peace Corps volunteer to teach blind students in Morocco. As a blind person himself, he is thrilled to be able to teach blind people in a foreign country. He learns Moroccan Arabic, meets new friends, and has a great time, while benefiting others.	MIUSA (see page 115) National Clearinghouse on Disability and Exchange (see page 116)	Discover the world in your own unique way Use your talents and connect with other disabled people worldwide Become enlightened by the cultural and linguistical diversity Practice language skills Prepare for an international career

As you can see from the different scenarios, international experience benefits everyone—students, professionals, career changers, retirees, disabled people—and sometimes in unforeseen ways. Knowing that there are international opportunities out there for everyone is the first step toward finding the right one for you.

Creating Your Path

Imagine it is a bright, sunny day and you are sailing on a lake. You see your destination on the other side of the lake and think it will be easy to simply sail straight on over. During my first sailing lesson, my Dutch sailing instructor kindly informed me that we couldn't sail "straight" over. We would have to "tack and jibe." Tack and jibe? Was I mistaken or was that an old '70s dance step? My sailing instructor quickly translated my new sailing lingo into words I could understand. Sailboats cannot sail in a straight line, therefore it is necessary to zigzag, or tack and jibe, to reach a certain point. When it seems like a sailboat is going off course, it is in fact, taking the easiest and most direct route to reach its destination.

For sailors, international adventurers, and global-career seekers alike, tacking and jibing is often the only way to reach your final global destination. Zigzagging your way through an international life is not impeded progress, but rather an ingenious way to weave your experiences into a solid and distinct life path.

How does this whole tacking-and-jibing business relate to your life? To acquire the skills you need to be successful you may have to go back and forth between your home country and various destinations in this seemingly indirect manner. Along the way, you will be picking up the skills you need from both the domestic front and the international market, weaving your global expertise with your domestic experience to create your international life path. Observe in the following example how Joe earned international experience as he zigzagged his way to a life overseas.

Your global life may also consist of journeying to various parts of the world, coming back to your home base, and then venturing out again, gaining more international skills and knowledge with every tack and jibe. If you already have academic, travel, or professional experience, you can start sketching your journey, using Joe's experience on the next page as a guideline.

International experience gives you endless opportunities for personal and professional growth and a more enriching life that you simply cannot get in your home country. One of the best things about your global ambition is that you have full control of how you plan and integrate overseas experiences into your life. No one forces you to learn a language, travel abroad, work in foreign countries, or help your global neighbor—it is all up to you. Whether you already have some international skills and experience, or if you are just starting to introduce yourself to life outside your country, you can learn more about your options and how to plan the best experience for yourself in the pages that follow. Good luck!

Joe wanted to be sent abroad to work as an expatriate. You can see the zigzag steps he took to develop his interests in South America, gain professional experience domestically, and then merge his interests to develop a unique global life for himself.

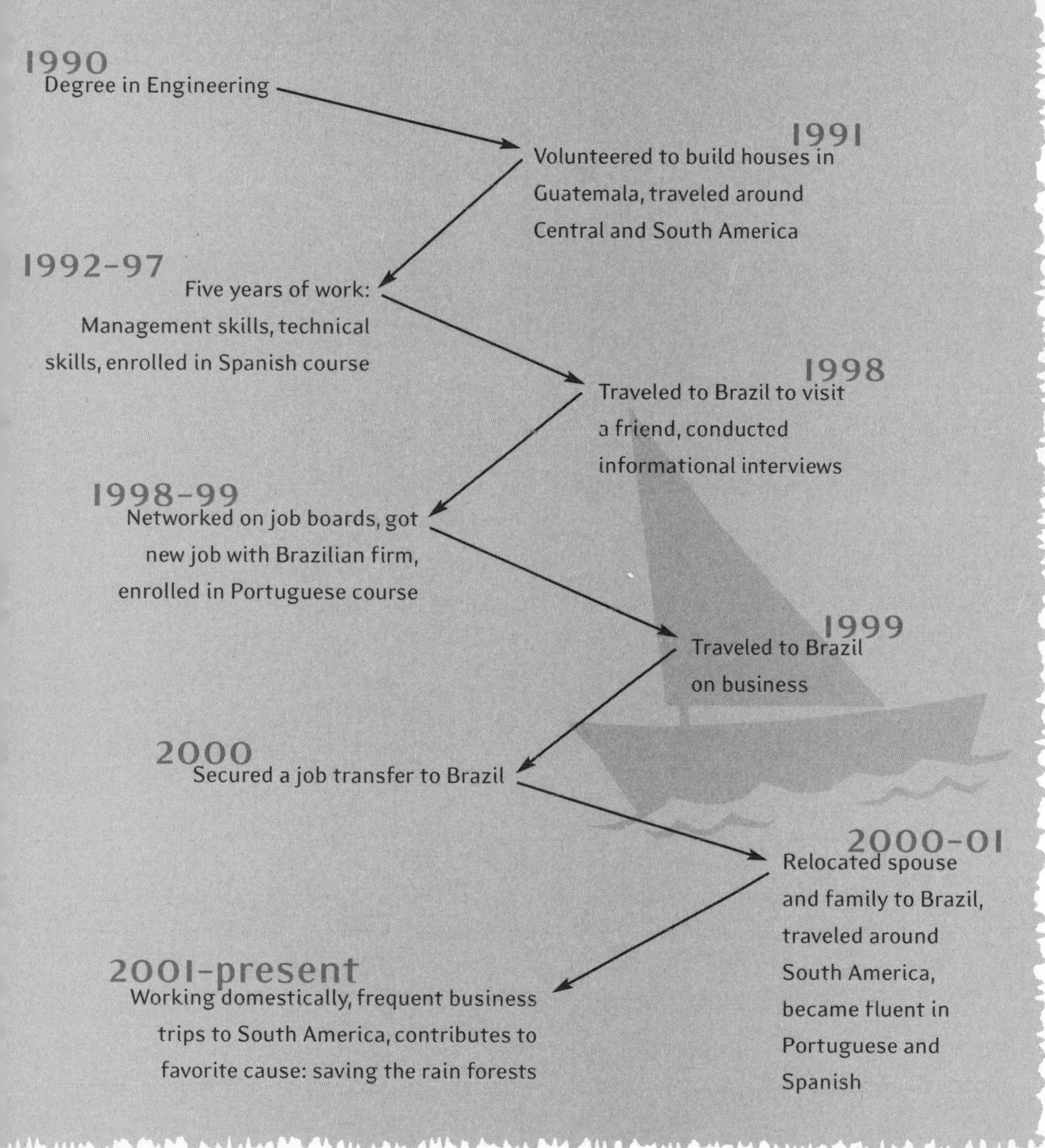

After living and working abroad, Joe learned the value of speaking foreign languages fluently, dealt with foreign cultures on a business and personal level, became involved with global issues that he never faced in his own country, and taught his children the importance of an international education. From this example you can see how over the course of ten years Joe was able to accumulate the twelve characteristics of a global citizen, enabling him to reach his personal and professional goals. Traveling, studying, volunteering, and working abroad are colorful threads that connected all of Joe's experiences, putting him on the international path in life.

YOUR JOURNEY

Domestic Experience

Education:

Travel/Interests:

Professional Experience:

International Experience

Education Abroad:

Travel/Interests Pursued Abroad:

Professional Experience Abroad:

PART ONE

How in the World Do I Get Started?

The fool wanders—the wise man travels.

THOMAS FULLER

In my experience helping worldly souls take advantage of overseas opportunities, I've found that although many people have short-term goals to travel, study, or work abroad, their long-term desires are to live meaningful and enriching lives. By knowing yourself, what drives your passions, and what your cross-cultural preferences are, you'll be well prepared to live a fulfilling life, both overseas and at home.

CHAPTER ONE

Uncovering Your Global Passion

This above all: to thine own self be true,
And it must follow, as the night the day,
Thou canst not be then false to any man.
WILLIAM SHAKESPEARE

More than likely your head already is buzzing with ideas about what you want to do when you go abroad. Are you dying to have an international career? Are you passionate about exploring the Mayan and Aztec Indian ruins in South and Central America? Or maybe you are a committed environmentalist ready to get involved in saving the world's rain forests? Being aware of your interests and passions is an important step to achieving a global life.

How do you get started on this exciting journey (or continue on an international path you began long ago)? It all begins with an introspective look at your life and some long-range planning. From there you will be amazed at the opportunities that can boost you into an exciting international life and career.

In my own travels abroad, I learned many skills that have helped me live a successful and rewarding life in various countries. I was at the beginning of my career when I first went overseas. I found that part of my education included taking a good look inside myself to clarify the direction I wanted my personal and professional life to go. I had to peel off several layers of myths, stereotypes, and social stigmas that I had grown up with and replace them with the newfound wisdom that would guide my international life.

For example, I had been taught that professional success is represented by a high salary, a prestigious position, and expensive material possessions. Although this definition of success didn't seem to fit with my values, the social pressure to "succeed" in this way was blocking my progress toward a fulfilling life. After much soul searching, I came up with a new definition of success, one that felt more comfortable. I decided that, for me, success means achieving my goals. At that stage of my life, I wanted to gain professional experience in various countries, become fluent in two languages, develop international friendships, and see as much of the world as possible. My mobile lifestyle didn't allow for the acquiring of many material possessions, climbing the corporate ladder, and donning the golden handcuffs that a high salary buys. Defining "success" is just one example of the many career- and life-planning issues I had to work out—and it's one I often see other aspiring internationalists struggle with. So, before you set off to travel, study, work, or play in the great unknown, let's make sure you're prepared for whatever's down the road.

Know Thyself

I've always considered myself to be slightly different from the American norm. Although I've always been able to "fit in" with most groups or organizations I wanted to be part of, I never felt 100 percent comfortable. After numerous self-assessments and many hours of introspection, I now realize that I was a moderately extroverted, entrepreneurial, and somewhat creative type trying to fit into a very extroverted, corporate, and rather image-conscious person's shoes. (If you've ever done a Myers-Briggs personality test, I am an ENFP—Extroverted, iNtuitive, Feeling, Perceiving type—but thought I was more Thinking and Judging—an ENTJ.) This realization has made a tremendous impact on my outlook and on my ability to successfully choose jobs, friends, groups, and opportunities that fit who I am. So what's my point? Know thyself. Know your natural preferences, your interests, and your passions. This knowledge will help you deal with people, it will help you in your career, and it will help you adapt to an international life. Here's the fun part: Once you decide who you are, you can venture out into the world and see if you are indeed who you think you are.

The easiest, cheapest, and most straightforward way to define your passions, interests, and likely direction is by taking a personality or career assessment. Many assessments can be found on the Internet or in the books listed below. Although these assessments may hint that, based on your personality and interests, you'd make a good teacher, consultant, doctor, or social worker, none of them will tell you that you are perfectly suited to follow an international path. As insightful as they are, these tests aren't set up to make a connection between an "international" personality, the associated qualities and skills of a global citizen, and the increasing number of traditional and nontraditional career options and life directions. Nevertheless, personality and career assessments are perfect tools for gaining a clearer understanding of your natural preferences, i.e., the things you really like to do. If you want assistance interpreting the results, career counselors can help. Find them at high schools, colleges and universities, and listed in the phone book as individual consultants.

My goal is to help you understand that there are innumerable dynamic and challenging learning opportunities and professional options for you overseas. However, they are simply inaccessible if you never venture from your home country. Becoming a global citizen is largely a lifestyle and career decision of your own making. Unfortunately, this is not something that all career counselors are aware of or encourage job seekers to pursue. This book will help fill the gap. I've tried to provide you with a template for an international approach to life, and the guidelines and resources to launch you in the right direction. If you ever need more answers than this book can provide, then it is time to find mentors in your field to lead you the rest of the way.

As helpful as people will be throughout your life, you'll discover that the important answers about achieving your life's ambitions lie within you. That is why knowing yourself and your passion is essential.

Assessment Resources

Take a few minutes to do an online assessment or invest in one of the recommended books listed below. At the very least, you'll gain a good grasp of who you are, what needs you have, and an outline of your personal and career goals.

CAREER ASSESSMENTS

★ **Job Hunter's Bible**

www.jobhuntersbible.com

This very helpful site by Richard N. Bolles, author of the best-selling *What Color Is Your Parachute?* (Ten Speed Press, updated yearly), offers links to career and personality analyses, along with other site referrals. Click on "Take an Interactive Test" to link to a page where you can choose various online career tests. Under "Career Tests," I liked the Princeton Review Career Quiz (from the developers of the Birkman method career test), which took only five minutes to complete. At the end of the twenty-four questions, you find out your job interests and your work style, based on your preferences. You can also see a list of careers that match your profile. I spent an additional fifteen minutes reviewing a career report that included general job strengths, job families, a job-titles summary, and my ideal work environment.

★ **What Color Is Your Parachute?**

by Richard N. Bolles (Ten Speed Press, updated yearly)

This book is updated yearly with all the latest and greatest job-hunting information around. It's an easy-to-read and humorous book, but most importantly, the job-hunting steps have been proven over the last three decades. That is why it is "the best-selling job-hunting book in the world."

PERSONALITY ASSESSMENTS

★ **2H**

www.2h.com

This site links to various personality tests available on the Web and includes the approximate amount of time required to take each test, ranging from five to twenty minutes. I liked the Keirsey Temperament Sorter II, which consisted of seventy multiple-choice questions and took only ten minutes to complete. It had an interesting feedback section that pinpointed my character and personality traits, and jobs that would be good matches for me. I was impressed with the accuracy. The Keirsey Temperament Sorter II is also available in twelve different languages.

The famous Myers-Briggs Type Indicator (MBTI), which analyzes your personality, thereby making it easier for you to find the right career based on your natural preferences, is not currently available online. I personally find the MBTI to be a valuable self-analysis tool—it's worth taking. You can take the abbreviated version of the MBTI in the books listed below, or see a career counselor to take the full assessment. (The two books analyze your career direction based on the MBTI.)

★ **DO WHAT YOU ARE: DISCOVER THE PERFECT CAREER FOR YOU THROUGH THE SECRETS OF PERSONALITY TYPE**
by Paul D. Tieger and Barbara Barron-Tieger (Little, Brown and Company, 1995)

This book helps you choose a career direction based on the Myers-Briggs Type Indicator (MBTI). Paul Tieger also hosts a message board on Monster.com, where you can do a quick quiz to analyze your personality and then post questions relating to your personality type and career.

★ **TYPE TALK AT WORK: HOW THE 16 PERSONALITY TYPES DETERMINE YOUR SUCCESS ON THE JOB**
by Otto Kroeger with Janet M. Thuesen (Dell Publishing, 1993)

The sixteen personality types are based on the Myers-Briggs Type Indicator (MBTI) and give an interesting and thorough analysis of how the use of personality typing can be applied to the workplace.

Figure Out What You Love (and Do) Best

The best advice that anyone has ever given me (and that I have passed on to others) is simply to "follow your passion." Do what comes naturally to you, do something that makes you feel fulfilled. At the end of a day's work you should feel energized; time should have passed by without you even noticing it. Develop and follow a path that allows you to live life the way you want.

OK, I know that's easier said than done—especially if you don't know yet what your passion is. That is why the personality and career assessments are important first steps to uncovering your hidden passions. Unfortunately, as a society we tend to follow well-trodden paths instead of striking out on our own and making the most of our talents, skills, and interests. As we grow up, our childhood dreams become massively cluttered and unconsciously rearranged with other people's viewpoints, our parents' hopes, and society's expectations. Maybe you always imagined you would be an artist in Paris or run a successful business in London, Bangkok, or Tokyo. Perhaps you saw yourself building schools in a rural Mexican village. As you learned more about the "real" world, maybe you convinced yourself that those early ambitions were too lofty or too unrealistic to get you where you thought you wanted to be.

The media's obsession in recent years with e-business, technology, dot coms, and the like can heavily influence our education and career paths. For security's sake and future career potential, you might feel like you should follow the hottest trends in information technology, finance, or marketing, rather than doing what you truly love. Don't get me wrong, global business offers many dynamic career opportunities, which is fine if you like business. But, if you are a budding artist, political activist, or freelance writer or photographer, working in the corporate world may go against your natural grain. The good news is that you can do anything you put your mind to with a positive attitude. And since you are reading this book, I assume you are ready to dust off those childhood dreams and start living the global life you've always wanted.

Pinpointing Your Global Direction

by Mark Thompson (Matin1066@aol.com)

I am a young attorney working in the Washington, D.C., area. I speak Portuguese and Spanish, and am searching for an opportunity to do something socially redeeming with my training. Most people have advised me to stay with the practice of law because they think law is a lucrative profession, and that's the only thing that matters. Perhaps if I believed that money alone could compensate for feeling like I'm not living up to my potential, I would indeed stay in it.

I went to law school, not out of any great personal conviction that this was my calling in life, but rather because my parents expected me to go. Now that I have paid my familial dues, so to speak, for the first time I am turning my efforts toward the pursuit of my own goals and life interests. Principal among these interests has always been language and history. My interest in Portuguese began when I was six and my father, a naval officer, was posted to the Portuguese island of Terceira. Later, I began studying Spanish in junior high, and by the time I reached the University of Virginia, I was a foreign affairs major studying Spanish and (Brazilian) Portuguese.

Before I entered law school, I had backpacked through Europe, lived in the Netherlands for a bit, and returned briefly to work in the Azores. I spent seven months working at the U.S. Embassy in Sudan and was teaching myself to read Italian. I was doing the kinds of things that I found fulfilling and interesting. I had entertained the idea of going to grad school to study Portuguese and history in some form or another, but my parents were hostile to the idea. In their minds, this was not a practical and useful way to spend my time and (their) money. So instead, I entered law school. At that time in my life I thought my parents had benefits of age and experience that I did not, and time would prove them right and me wrong. Well, I am perhaps the only student in the history of Duke Law School who smuggled a book on Portuguese history (in Portuguese) to class to escape the grinding boredom of criminal law. I gave my parents' goals more than a fair shot, and the effort ended up displacing all of my other interests in life. Now that I have decided to leave law, I am reevaluating whether or not to pick up where I left off before law school.

I decided that for me, job satisfaction would be found with a job that connects to my own interests and talents, but first I had to reconnect with those interests and talents—they had remained unused and unexamined for about five years. I went through my belongings and discarded about one-third of all my stuff. Organizing photos was very instructive. I observed that I looked tired and stressed out in most of my photos from law school, but the photos I have from Africa show a much happier me. As I went through my belongings, tossing them into "keep" or "throw-out" piles, I began to reconnect with my interests. I began to perceive the emergence of an overall pattern: The stuff I was most eager to toss was related to law school, and the stuff that I kept was related to my interests in history, languages, travel, and sports.

In addition to editing my personal belongings, I set about renewing the lateral social ties that I had neglected during the time that my life focus was centered exclusively on law. I rebuilt contacts with friends, and as I did so, in each case I asked myself: "Why am I choosing to reconnect with these people? What value do they add to my life? What value do I add to theirs?" I found that these were people who had originally chosen to interact with me because of who I was, not because of my career choice. These were also people who were generally happier with their lives because they had followed their own interests rather than fulfilled the expectations of other people.

At the same time, I also reestablished my vertical network, reconnecting with people who had been my mentors before I went to law school. They gave me perspective on my current problems and provided insight as to how I might resolve them. I also asked them: "Why do you do what you do? If you could do it all over again, would you still choose the same profession?" Understanding how other people feel about their own career paths and how they link to their interests and personalities helps to provide insight into one's own questions in these areas.

And of course, there was the immediate financial problem. In order to lighten my financial load, I restructured my school loans, found a roommate, and consolidated my credit card debt.

I also reevaluated my "marketing strategy." I redrafted my résumé to emphasize transferable skills such as overseas experience, management, research, and analytical and writing ability. I downplayed the attorney bit, since I had no intention of returning to another law firm. I learned HTML and increased my proficiency with various types of office computing software. I attended night classes, enrolling in an e-commerce certificate program at a local university, which, aside from teaching business skills, also exposed me to contacts outside of the legal field. Through a temp agency I took an interactive CD-ROM-based Spanish proficiency test. If any employer has doubts about my ability to speak Spanish well, I can simply produce the results of my Spanish test as proof. I even took a bartending course. Bartenders seem to find jobs anywhere in the world, and the initial time and money investments for training and certification are very low. (I am now the resident "gringo poliglot" bartender at a Spanish tapas bar.)

It is likely that my next step may well be to return overseas. I have applied to the Peace Corps for a position as an English teacher, which would most likely land me somewhere in Eastern Europe. However, I have also made inquiries with individuals involved in the reconstruction of East Timor, which had been a Portuguese colony before its thirty-year occupation by Indonesian forces. The focus of my pitch is that I have experience living in difficult locations and can teach English, Portuguese, and the basic fundamentals of law—all of which are greatly needed in East Timor right now.

Even though it is difficult to uproot oneself and walk away from the seductively comfortable routine of American life in order to face the uncertainties and (quite often) physical dangers presented by life overseas, the compensation in terms of personal development is worth it. It is easy to make life and career decisions based solely on financial concerns and to conform your life to the contours of whatever job will pay the most money. That is what most of us Americans are culturally programmed to do. However, if you put the money factor aside, shift your mental frame of reference, and instead analyze your life in terms of the plot of a novel, the results of your analysis will most likely change. Imagine that you are stuck on a long train ride and must choose one of two books to read in order to pass the time: the first is a novel whose main character is an office worker who is essentially working to pay his monthly cable bill; the second is about someone who decides to travel in South America (and of course encounters various setbacks in the process), but who pushes beyond the boundaries of conventional American life. Which would book would you pick up to read? Indeed, which of the two characters would you rather *be*?

Truthfully, I'm not sure where I'll be going next, or what I'll be doing when I get there—but this uncertainty isn't necessarily a bad thing. After all, uncertainty is an inherent component of risk, and I'd much rather be a risk taker than an office drone who's just working to pay the cable man at the end of the month.

Read more about Mark Thompson's international experiences in chapter two.

Living abroad gives you a whole new perspective on life and helps you decide what you value most. For many people, some aspect of international experience is their passion: business, languages, travel, cultural anthropology, helping people in developing countries. Your passions might be even more exotic, like climbing the world's tallest mountains, diving the world's most beautiful reefs, photographing the world's most spectacular wildlife. We rarely hear about the exciting lives of people following their international dreams—like the adventure tour guide in Belize, the *National Geographic* photographer in Africa, the Amnesty International activist fighting for human rights in China, or the educational trainer delivering cross-cultural courses to international companies. What about the many ESL (English as a Second Language) teachers, the educators working in international exchange organizations, the owners of language school franchises, expatriate spouses who make an independent living consulting and writing books, artists who sell their works to international hotel chains and galleries, and retirees who travel in their golden years? We don't hear these stories because these people are too busy out living out their dreams to tell us.

Although many of these folks may not be wealthy by American standards, they are fantastically rich in worldly experience and life enjoyment, rich beyond what many of us dream is possible. A life path doesn't include just the work you do to earn money, it encompasses everything from where you work, to the people you choose to associate with, the hobbies you are involved in, and the causes you support. All of these are influenced by what you value most.

Defining Your Passion

Experts agree that anyone can live a successful and satisfying life by serving others in a way that makes use of their unique talents, passions, and skills. Knowing what you like to do and what you do well will help you carve out a niche for your skills. Whether you call this your mission, your passion, your purpose, or your life's direction, it all comes down to knowing how to best combine your skills and interests to serve the world.

Try these exercises for narrowing down your interests.

- Start with something you like to do and dig deeper and deeper until you decide it is a passion or until you want to try something else. Whether your interests are music, art, the stock market, or sports—keep reading, talk to people, and stay involved in your field.

- Expose yourself to different types of people, activities, and mentalities. Do things you wouldn't ordinarily do—see a foreign film, join an ethnic club, or take a foreign-language course.

- Gain a comparative perspective by pursuing your interests in a foreign country. Having taken that first step and lived overseas, you'll have a clearer understanding of your interests and priorities in life.

- Visualize the location you see yourself living in overseas. Are you on some remote island teaching tourists how to scuba dive? Are you running an international office in Beijing? Are you leading expeditions in the Arctic?

- Cut out pictures of your dream life and make a collage of the things you love. Look for patterns that define who you really are as a person.

Answer these questions to help you sort out your goals. (I included my answers to show you how simple your answers can be.)

- What is your favorite thing to do?
  (I love living in foreign countries, learning foreign languages, exploring the world, and helping others do the same.)

- What skills do you most enjoy using?
  (I enjoy writing, strategic marketing and sales, researching, and foreign languages.)

- Where do you want to live and work? What is your working environment like? (To fit my lifestyle, I envision working from my laptop from anywhere in the world, at hours that are convenient for me.)

- Who do you want to help or serve?
  (I want to help people, especially the younger generation, reap the rewards of traveling , studying, working, and living overseas.)

- How do you think you could combine your skills to serve others?
  (I want to help people enjoy the rewards of their international experiences. I prefer to communicate my message in writing, through books, articles, etc.)

After that bit of self-analysis (and a few more days—weeks, months, years?—of deep introspection), you should have a better idea of how you want to develop your life. But are you culturally prepared for the countries you want to visit? You'll find out in the next chapter.

CHAPTER TWO

Culture Prep: A Mini-Course for the Culturally Challenged

A mind that has been stretched by a new experience can never go back to its old dimensions.

OLIVER WENDELL HOLMES

Although some people say that going abroad is not for everyone, I believe everyone who goes abroad will benefit in some way. The key is to decide ahead of time what you are comfortable with and how much you are willing to be culturally challenged. The answers to these questions depend on your values, your personality, and your goals.

Cross-Cultural Assessment

Leaving the safe environment you've known and been influenced by since birth and going to a place where life is different will challenge you to think seriously about what is important to you. Can you deal with buses and trains that never show up on time, as is often the case in South American countries? Can you put up with garbage, litter, and maybe even animals, in the streets, like in India? Can you accept a culture where women take on more traditional roles than in your society? Understanding the typical characteristics of different cultures is essential to recognizing your own cultural tolerance.

Cultures are generally classified in very different ways: The Swiss are punctual and detail oriented. The polite and respectful Japanese must never lose face. Risk-taking Americans are driven by money and profit. Although you might be aware of some of the typical cultural classifications, do you know your own cultural profile? Surprisingly, it may not necessarily jibe with the cultural profile of your home country. So, which cultures are most similar to your natural preferences? Which cultures are vastly different? Many global citizens, by the very nature of our interests in other cultures, deviate from the cultural norm of our societies to some extent. In the cultural assessment below, my natural tendencies are much closer to the multi-active cultures of southern Europe than to the linear-active culture of my native country, the U.S.A. Finding out where you stand, culturally speaking, will help you understand yourself better compared to the culture in which you choose to live, travel, or work. And it will give you an overview of the similarities and differences of various other cultures.

The Lewis Model of Cultural Classification

The following survey and diagrams were created by Richard Lewis Communications, Plc. (www.crossculture.com), a cross-cultural training organization known worldwide for its forty years of extensive research and training in cross-cultural issues. The model is based on Richard Lewis's own experience speaking twelve languages and being a pioneer in cross-cultural issues around the globe. Classifying cultures gives us a general framework for understanding cultural behaviors. Of course, individual behaviors will differ based on circumstances, but the general conclusions that the cross-cultural classification provides are invaluable tools for helping people live and work more successfully across cultures.

There are three cultural categories of the Lewis Model: linear-active, multi-active, and reactive. You'll get a brief introduction to each category as you work through the fifteen questions in the Cultural Classification; a more detailed description is given after the survey.

HOW TO FIND YOUR CULTURAL CLASSIFICATION ON THE MAP OF WORLD CULTURES

1. Find your cultural profile by answering the fifteen-question survey in figure 1. Circle the description that best describes you. Add up the number of answers circled in each category and enter the total at the bottom of the corresponding column.
2. Next, transfer the totals for each column (linear-active—L, multi-active—M, and reactive—R) to the LMR box in figure 2. A sample, filled-in chart is shown in figure 3.
3. Using figure 3 as a model, plot your LMR score on the numbered lines in the triangular diagram. Each category is represented by three numbered lines. One represents the low end of the spectrum, while fourteen represents the high end. As you can see in figure 3, which shows an L score of 11, an M score of 2, and an R score of 2, the number 11 is circled on the linear-active line, and number 2 is circled on both the multi-active and reactive lines.
4. After you fill in and circle your LMR score, connect the dots as shown to form a triangle. Place a dot in the middle of the triangle. This dot represents where your natural preferences stand in relation to the cultures of the world. You can then draw arrows to the cultures you most closely relate to as well as the cultures you vastly differ from. In figure 3, you can see that the sample person's cultural preferences correspond closely with those of the Netherlands and Norway, and are furthest from Hispanic America, Arab countries, and China.

Figure 1

	LINEAR-ACTIVE	MULTI-ACTIVE	REACTIVE
1	talks half the time	talks most of the time	listens most of the time
2	plans ahead step by step	plans grand outline only	looks at general principles
3	polite but direct	emotional	polite, indirect
4	uses official channels	seeks out top/key person	uses connections
5	partly hides feelings	shows feelings	hides feelings
6	does one thing at a time	does several things at once	reacts to partner's action
7	dislikes losing face	has good excuses	must not lose face
8	job-oriented	people-oriented	very people-oriented
9	confronts logically	confronts emotionally	never confronts
10	rarely interrupts	often interrupts	does not interrupt
11	truth before diplomacy	flexible truth	diplomacy before truth
12	sometimes impatient	impatient	patient
13	limited body language	unlimited body language	subtle body language
14	uses mainly facts	puts feelings before facts	statements are promises
15	separates social/ professional	mixes social/ professional	connects social/ professional
TOTAL			

Figure 2

WRITE YOUR LMR SCORE HERE:

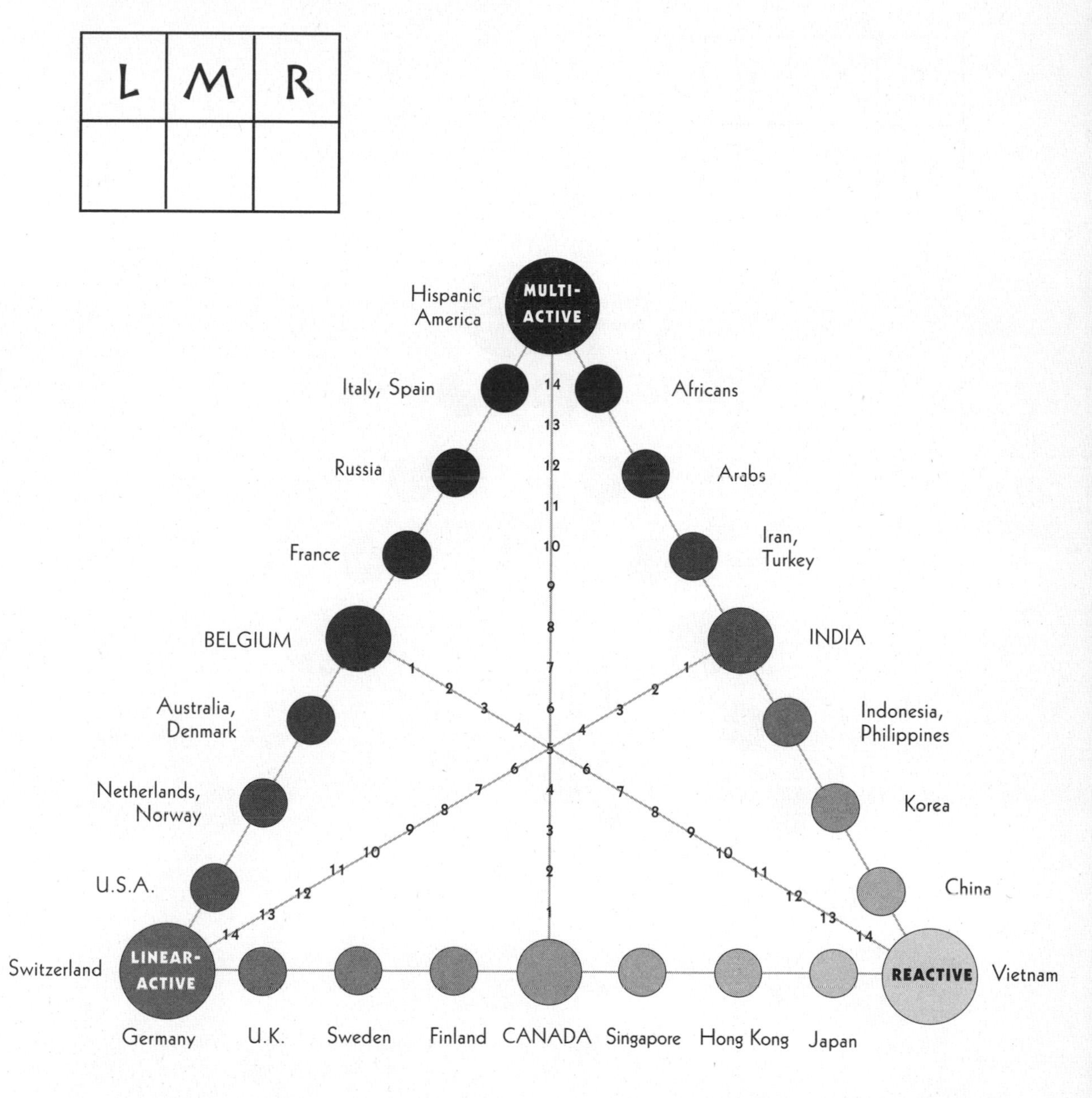

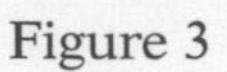
Figure 3

EXAMPLE:

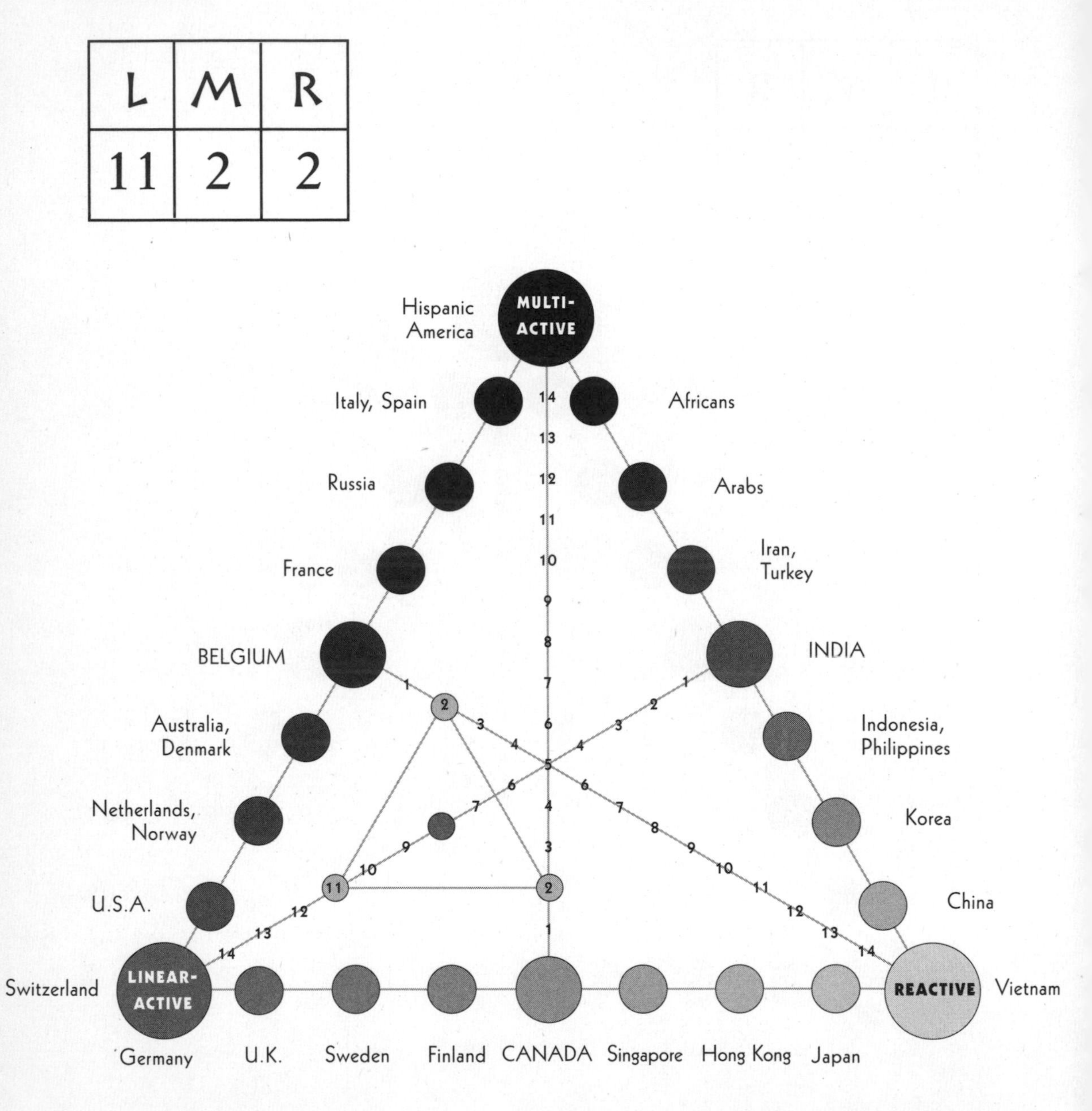

L	M	R
11	2	2

Lewis Model Summary

- How close is your cultural profile to those countries you plan to visit?
- How much are you prepared to adapt your thinking and behavior?
- Even with cultures that are similar to your profile, do you know what the differences are?

Linear-Active Cultures

In addition to the characteristics listed on the cultural classification, linear-active cultures tend to be data and fact oriented. Speech is mainly used to exchange information. Linear-active cultures tend to stick to planned agendas, work fixed hours, be results oriented, and gain status through achievement. Linear-active people will secure deals by compromising. Contracts are binding and short-term profit is most desirable. In general, linear-actives rarely borrow or give. The cultures at the high-number end of this scale are Germany, Switzerland, the U.S.A., the United Kingdom, Sweden, Finland, and Norway.

Multi-Active Cultures

Multi-active cultures are dialogue oriented. In general, the people are animated, they love to talk (usually quite fast), they show their emotions, and they value relationships. Multi-actives like a flexible work schedule and sometimes they ignore the rules. They would rather win an argument than compromise. Because relationships are more important than fixed appointments, they tend to complete meetings with friends or colleagues before moving on to the next meeting, even if that means showing up later than the scheduled time. Multi-actives achieve status in their culture by having charismatic personalities and connections. In business, having a good relationship with the client is deemed more important than drawing up a contract, which is considered more idealistic than realistic. Pulling strings and doing favors is a way of business and life. Mexicans, Colombians, Italians, Greeks, and Spaniards are at the heart of the multi-active scale, followed by some African and Arab cultures.

Reactive Cultures

Reactive cultures are the polite listeners of the cultural categories. They prefer a slower-paced dialogue to have time to reflect on what is being said, they are sometimes suspicious of fast talkers. They never confront as they must never lose face. In addition, reactive cultures prefer to live in harmony with people and the environment. Speech is a reflection of their desire to promote harmony in relationships, as is the practice of compromising. Reactive cultures are respectful and give ritually, but rarely borrow. Work, free time, and life are intertwined. A person gains status in this society through birthright and education. In business, long-term profit and increased market share are important. Ancient philosophies, like Confucianism, dominate the way of life. Reactive cultures are found typically in Asia, with China, Japan, and Vietnam being at the far end of the spectrum.

CULTURAL PREPARATIONS

Did you know it is a major faux pas to use the familiar "you" with your Swiss or German boss? Or that it is forbidden to eat or shake with one's left hand in Arab countries? And that it is taboo to say "no" or "can't do" in Japan? You might think this cross-cultural stuff isn't that complicated, when in fact it can take many years to get into the mind of the culture you are visiting. More important than being aware of which cultural blunders you should avoid, it is essential to understand why people think and act the way they do. As you uncover the cultural justification behind the behaviors of different nations, you'll see how years of history, politics, religion, social customs, traditions, and language have produced the current norms. By understanding the values that lie behind the cultural behavior, you'll be much better prepared to handle a variety of cultural situations, not just avoid the typical foreigner mistakes.

Despite your best efforts to grasp cultural differences, some things will just be beyond your comprehension. Not because you aren't enlightened, but because some cultures have customs that go back hundreds, if not thousands, of years and simply can't be fully understood quickly. If cross-cultural communication was really that simple, million-dollar deals between foreign companies wouldn't fall through as often as they do. Unfortunately, many companies still do not invest in cross-cultural training to equip their overseas personnel with the skills necessary to interact, lead, and negotiate successfully. When it comes to cultural training, you might be left to your own devices. A good guidebook of the country you plan to live in can give you some tips and pointers, and a cross-cultural seminar will give you more in-depth cultural preparation. However, nothing beats the education you'll get from countless cultural mistakes you are bound to make while living abroad.

The potential for making embarrassing cultural mistakes will be present at every turn. If you are walking down the street minding your own business, and take out a tissue to blow your nose, you may end up offending somebody—that is if you are in Japan, where nose blowing in public is a big no-no. It is these common everyday acts that will call attention to your inappropriate and foreign behavior. You might sense you did something that wasn't quite kosher, but you won't know what it was. Unless you've had some cultural training, you'll repeatedly make the same mistakes in different contexts until your cultural blunder dawns on you or a local tips you off. On the other hand, if a person from another culture tries to politely accommodate you and your culture by speaking your language or adjusting to what they think are your norms, the cultural blunder may come from their side.

Here are a few cross-cultural situations in which you might find yourself.

Greeting Protocol

Meeting people in a foreign country usually involves some sort of protocol, starting with language. Many countries have a formal greeting that is used between people who are not close friends. The French, the Germans, and the Japanese use the formal "you" according to a person's social position, rank, or title. The informal and "friendly" American approach to meeting people, reflected by the use of first names, a wide smile, a vigorous handshake, and maybe even a slap on the back, is irritating, offensive, and unprofessional in many countries.

In business, the exchange of business cards can be a formality in itself. Try not to start off your new business relationship on the wrong foot by quickly glancing at your counterpart's business card and then shoving it in your pocket. Learn to show respect by accepting the card with two hands. Read it thoroughly, carefully noting the person's title, and learn to properly pronounce their name.

One Kiss, Two Kisses, Three Kisses, or Shake?

by Elizabeth Kruempelmann (ekruempe@hotmail.com)

My German husband and I are always trying to master the proper protocol for greeting our many foreign friends. We have been living in Portugal for several years, so we've become quite used to the Portuguese kissing routine. Two kisses upon greeting and departing from friends is the norm. However, there is a proper kissing procedure that you must know if you want to do it right. Touch cheek to cheek, starting with the right, and lightly kiss the air. If you do it any other way you might make the procedure a bit more intimate than it is supposed to be!

If you are used to one greeting protocol, switching gears to a different one can take a bit of adjusting. When we return to my husband's country for visits, I attempt to kiss all of our German friends out of habit and inevitably cause a bit of confusion. Comfortable with a reserved distance between people, our German friends usually become momentarily flustered by an American trying to kiss them twice. In Germany, a simple handshake, if anything at all, will do. Noticing my misplaced protocol midway through the process and attempting to revert to the German handshake usually complicates the greeting even more. The result is a rather awkward, yet comical display of the kissing protocol gone astray, ending in an exchange of slightly painful head bumping and harmless nose clashing.

After a few days of handshaking in Germany, we return to Portugal where we meet up with our Polish and Belgian friends who are living there. However, instead of the customary two Portuguese pecks, the Poles and Belgians give three kisses (right cheek, left cheek, right cheek). And if we forget the third one, the whole thing once again turns a bit strange and comical for any spectators who might be watching this rather odd intercultural greeting game.

Read more about Elizabeth Kruempelmann's international experiences elsewhere in this chapter as well as in chapters five, seven, and eight.

Find out the proper formal or informal greeting that should be used in the country you are visiting. It may be accompanied by kissing, shaking hands, bowing, or some other courtesy. Learning how to alter your body language to make a good first impression will work to your advantage.

Body Language

There is verbal language and then there is body language. Both are used to communicate our intentions. Body language encompasses an invisible space bubble surrounding our bodies, our facial expressions, and our physical gestures. Taken together with its verbal companion, if you aren't careful, body language can make you the unassuming target of a cross-cultural disaster. The potential for sending out the wrong messages and seriously confusing those with whom you are trying to communicate is great. The proper solutions to this cultural quandary depend very much on the culture and situation. For example, what is a good sign that a business person from a multi-active culture like Mexico is ready to talk business? He will move closer to you, within a foot or so, closing in on the space bubble. If you are from a linear-active or reactive culture, you might feel your personal space is being violated and automatically back off. Big mistake! You've just unwittingly sent your potential business partner a nonverbal message that you are not ready to do business with him.

Facial expressions can also be confusing. To a Swede, a smile is a sign that things are going well. Yet when things are going equally well for a Finn, there may be no trace of a smile at all. The American smile, though meant sincerely, is often seen by other cultures as insincere and superficial. The Japanese, by contrast, smile to be polite or when they are angry.

Eye contact, or lack of it, can also be very important in communicating. However, the same amount of eye contact that is usual in Greece would be considered inappropriate, and even rude, in less expressive cultures like Japan.

Similarly, gestures in different cultures can take on a variety of meanings. Multi-active cultures, like the Latinos and Arabs, are typically characterized by an open display of emotions, they talk with their hands, getting their arms, head, and whole body involved in the communication process. Some linear-actives, like Scandinavian and Germanic cultures, might find multi-active body language rather distracting, annoying, and even misplaced in a business atmosphere.

As you can see, nonverbal language is indeed a form of communication. Like a foreign language, it needs to be learned to effectively interpret your new environment and make yourself understood.

Verbal Language

Your nonverbal chatter and the verbal language you use (or don't use) have the potential to either further confuse your listener or help clarify your message. What does it mean when your business presentation to a company in Helsinki is met with complete silence and stillness? It might be unnerving if you don't have a sense of the verbal and nonverbal feedback your audience is giving you. In fact, silence and stillness are forms of communication in Finnish society, where silence is golden. Chances are, very subtle body language and communication between the participants has slipped by your untrained senses. The restrained use of verbal and nonverbal communication is part of the Finnish culture and many other reactive cultures of the world like the Japanese and Chinese.

In contrast to the Finns, multi-active cultures love to combine their verbal expression with animated gestures in an attempt to clarify their message. If you regard this obsession with talking as nonsensical chattiness, you may be setting yourself up for failure. Talking is a way to build the oh-so-important relationship base that you will need to survive in southern Europe and South and Central America. The biggest cultural mistake you can make in these cultures is not to engage in small talk.

When you have people from all the cultural categories talking in nonnative English about things like "reasonable" delivery dates, "fair" contracts, and "common sense," you can imagine the confusion. These words can be interpreted in dramatically different ways. For instance, a "reasonable" delivery date in Portugal is not necessarily the date stated in the contract if conditions make it difficult to meet that delivery date. Multi-active cultures like the Portuguese need to be flexible. As a linear-active American you might argue that the contract was "fair" and final, despite the change of conditions. Well, similar to the multi-actives, the Chinese, Japanese, and

Language Blunders

by Elizabeth Kruempelmann (ekruempe@hotmail.com)

Verbal language blunders can be a two-sided game when a foreign language is involved. Either you are in danger of saying something offensive, off-color, or otherwise embarrassing when speaking a foreign language, or the other person could be put in a precarious position by trying to speak your language. The latter happened when a female Polish colleague and I were meeting for the first time with the Polish marketing director of a well-known Polish company.

When we arrived at the director's office, the director, who I'm guessing was in his early 50s, received us very graciously with typical Polish introductions. It was the middle of winter so both of us were dressed in heavy coats and scarves. As he led us to his private office, we reminded him that the meeting would be in English. He quickly pointed to a few hooks on the wall and responded with heavily accented English, "Yes, yes of course. Please, take off your clothes. I help you." I shot my Polish colleague a wide-eyed look of fear, despairing that we had somehow ended up at the wrong company. Also stunned and taken off guard, she quickly assessed the situation and returned a half-reassuring look that said, "I think he meant COATS, not clothes!" Phew! I'm not sure if it was a language blunder or a Freudian slip, but I like to think it was an innocent mistake. Needless to say, the marketing director's language faux pas, which I don't believe he even noticed, started the meeting on a rather awkward note. And it was difficult for my colleague and me to concentrate on our presentation without bursting out in silly laughter.

Read more about Elizabeth Kruempelmann's international experiences elsewhere in this chapter as well as in chapters five, seven, and eight.

A Global Citizen's Perspective

On Cultural Relativism and the Sudanese House Pet

by Mark Thompson (matin1066@aol.com)

One blazing-hot afternoon in mid-June, I met up with Saddiq, who was a student of dentistry at the University of Khartoum. Sitting in a bus shelter at the edge of campus waiting for his cousin to pick us up and drop us off at Saddiq's parents' house, we somehow found ourselves on the subject of household pets.

Dogs, in particular.

I inquired of Saddiq why it was that the Sudanese didn't particularly care for dogs.

Saddiq objected strongly to the assumption that lay behind my question, stating that the Sudanese loved dogs and often kept them in their homes as pets.

Surprised because this contradicted both what I had heard from other people, as well as what I had observed directly, I commented that I hadn't known this was the case and promptly apologized for my ignorance.

Graciously Saddiq accepted my apology and added that his family had always kept dogs around the house, and that he had always been quite fond of them.

Reflecting on how much I missed my dog, who was staying with my parents in Maryland while I was in Africa, I felt that we were connecting on a significant level—closing a cultural gap of understanding, if you will. Wanting to follow up on this moment of bonding, I asked Saddiq what his favorite dog had been called.

Blank surprise flashed across Saddiq's face: "You mean you Americans give your dogs names?"

Read more about Mark Thompson's international experiences in chapter one.

other Asians believe it is common sense not to be bound by a contract that was simply meant as a guideline in the first place, especially when market conditions change.

These are just a few of the ways your best efforts can take a turn for the worse if you are not culturally prepared. With the potential for cross-cultural misunderstanding quite high, you can see that cross-cultural awareness and training is essential for success.

Cultural Mini-Course for the Businessperson

Business people will undoubtedly come across challenging cultural differences in many areas. Here are a few examples of cross-cultural differences in the business world.

TIME

Let's say you are an American and are meeting a German in Frankfurt for a business meeting at 2:00 P.M. (or 14:00). What can you expect? This is an easy one. Both of you are from linear-active cultures where you segment your day according to time. Time is money and you will expect the other person to show up at the agreed-upon time of the meeting. Even if you are meeting a friend at a bar or attending a party, in Germany you are expected to arrive at the stated hour. It would be rude to show up late.

However, if you are meeting a Brazilian businessperson, don't be surprised if she arrives fifteen minutes to several hours later than the planned meeting time. Meeting times take on a relative meaning in Latin cultures. For Brazilians and other multi-active cultures, time is not scheduled according to the linear-active clock. Instead, multi-actives invest their time in the human aspects of meetings and relationships. Time spent building relationships almost always takes priority over being "on time" to something else. As a multi-active culture, the Brazilians prize relationships and will completely finish talking, socializing, and doing business before moving on to the next appointment. Assuming that they like you and that you take time to ask questions and get to know the personal side of your shared business, you'll be granted the same amount of attention as you would in other relationships. If you are invited to a party in Brazil and show up on time, you might have to pitch in with the preparations, as your hosts most likely won't be ready for anyone to show up for another half hour to an hour. As is often the case with parties in Latin countries, friends will pop in and out at various times throughout the event.

In contrast, showing up a few minutes early to a meeting in China, you might be surprised to see some or most of the participants already politely awaiting your arrival. Reactive cultures hope that by starting the meeting when you arrive, they won't take up too much of the time you scheduled for the meeting. Although the Chinese may make a point to thank you for your time and tell you early in the meeting that they do not want to take up a lot of your time, like the multi-actives, they do expect a generous amount of time to be invested in building long-term relationships.

ARE THEY LISTENING?

Let's say you are preparing to present your product. If you are an American, you'll probably create a savvy presentation that includes all of your product's bells and whistles. You'll entertain your audience with a few jokes and impressive company and product information. You know you have to go for the hard sell because you are up against some tough competitors. Time is of the essence, so your presentation will be short and sweet—thirty minutes max—but lively and to the point.

If your audience is American, they will likely listen carefully, be impressed by your product knowledge, and be won over by your humor. If you give this same presentation in the U.K. or Australia, chances are good that it will be a success. However, if the audience is composed of Germans, Japanese, or people from the Mediterranean cultures, they will be greatly unimpressed. Why? Because you have taken for granted the listening habits of your audience. Your presentation must be adjusted to the listening habits, attention span, and linguistic requirements of the target culture to achieve cross-cultural success. Too many companies never give their standard presentations a second thought, assuming that what works here must also work there. That is the first, and often most deadly, cultural mistake a company can make.

As a linear-active culture, the Germans (as well as the Swiss and Austrians) like facts, and lots of them. In fact, about 75 percent of their attention during a meeting is focused on listening for information about the reputation of the company, the quality of the product, technical points, and delivery details. Germanic cultures can listen for an hour or more about details Americans would find rather boring and would prefer to gloss over in ten to fifteen minutes. Germans do not necessarily want or expect to be entertained during a serious meeting. Humor and light-heartedness could irritate the German who wants to get down to business. The other 25 percent of the time, German businesspeople are listening carefully for pricing information and context. They also expect that information will be included in a packet of very detailed, printed materials about your company and its product.

To the Japanese, your presentation failed too, but for completely different reasons. One reason is that your presentation, complete with English clichés, slogans, idioms, and catch phrases delivered at a fast pace, was simply incoherent. As a reactive culture, the Japanese would listen to be polite and create a harmonious atmosphere, and would never lose face by admitting they didn't understand what was said. However, more than half of what you want to communicate will generally need to be explained patiently several times in various contexts for the Japanese to really grasp it. If you see several or even all of the Japanese participants close their eyes while you talk, this might well be a good sign that they are listening intently to understand you and not falling asleep as it might appear. It is important to hand out a printed version of your presentation, with diagrams and easy-to-understand language. Like the Germans, the Japanese don't expect to be entertained and prefer a rather quiet presentation over a lively one. Showing respect for the Japanese company and a formal presentation that promotes harmony and politeness is valued most. The Japanese are uncomfortable with any presentation that is too forward or that promotes confrontation or the butting of heads. They are patient, and like the Germans, prefer a presentation of at least sixty minutes.

Mediterranean cultures like the French, Spanish, Italians, and some Arab cultures expect a presentation on the opposite end of the scale from the Germans or Japanese. As multi-active cultures with short attention spans, they expect a short, lively, personal, nondetail-oriented presentation. They only need a general outline of the product. While you are talking away, your Mediterranean and Arab counterparts could be thinking of several other things while at the same time watching your body language. They are half listening for information while hoping more for imaginative ideas, eloquence, and a personal touch. Many times multi-active cultures will interrupt your presentation and take the discussion in another direction. Even with a short presentation of fifteen to thirty minutes, expect to spend more time talking afterward, establishing solid personal and business relationships.

I hope this cross-cultural mini-course has given you some insights into the various meanings behind common, everyday behaviors in different cultures. By identifying your own cultural profile, you'll be better prepared for the cultural challenges you'll undoubtedly face.

Note: Many of the examples listed above were pulled from Richard Lewis's monthly *Cross-Cultural Letter to International Managers,* in which the "concepts of time," "national listening habits," and many other cross-cultural topics are explained. The *Cross-Cultural Letter* is published monthly by The Institute of Cross-Cultural Communication (a division of Richard Lewis Communications Plc.), Riversdown House, Warnford, Hampshire SO32 3LH, U.K., Phone: 44 (0) 1962 771615, Fax: 44 (0) 1962 771355, Email: info@crossculture.com.

CROSS-CULTURAL RESOURCES

★ **THE CROSS-CULTURAL ASSESSOR®**

from Richard Lewis Communications, Plc., and Promentor Solutions Oy

The Cross-Cultural Assessor® is a unique interactive software tool providing one of the most advanced cross-cultural analyses for individuals and organizations. By completing a series of questions regarding your personal cultural style, cultural beliefs, and knowledge of world cultures, you will receive immediate feedback on your level of cultural awareness. The personalized nature of the program allows users to develop cultural skills, review a self-study section of over forty cultures, and learn about many cultural issues that will affect your experience overseas. For pricing and ordering information for individuals, organizations, and businesses, see www.crossculture.com, www.cultureactive.com, or call Richard Lewis Communications at 44 (0) 1962 771615.

Dos and Taboos Around the World for Women in Business
edited by Roger E. Axtell (John Wiley & Sons, 1997)

This book is designed especially for the increasing number of women who are working abroad. Cultural survival skills for women can be quite different from those for men. This book discusses cultural "to dos" and "not to dos," everything from climbing the corporate ladder to dating a foreigner.

Kiss, Bow, or Shake Hands: How to Do Business in 60 Countries
by Terri Morrison, Wayne A. Conaway, and George A. Borden
(Adams Media Corporation, 1995)

Covering sixty countries, this book will teach you the ins and outs of intercultural relations, like making appointments, business entertaining, greetings, forms of address, gestures, and proper dress.

When Cultures Collide: Managing Successfully Across Cultures
by Richard Lewis (Nicholas Brealey Publishing, 1999)

If you are going to live and work overseas, this book is essential for successful cross-cultural communication. It is an entertaining yet instructive look at a subject many people living overseas could stand to learn more about: how to successfully do business across cultures. A very thorough behind-the-scenes look at forty countries gives insightful guidance on how other cultures make decisions, value time, negotiate, interpret contracts, and more.

The Culture Shock Phenomenon

Cross-cultural preparation is an important step toward making your transition to a foreign country as smooth as possible. Knowing the proper dos and taboos is one thing, but putting that knowledge to practice can be another thing altogether. As you settle into your new home, you'll no doubt encounter some cultural bumps in the road. Understanding the local mind-set takes time. Some people spend months on an emotional roller coaster until they adjust to the cultural differences. Welcome to culture shock! It happens to most of us, especially when living overseas for the first time. Here's a look at what to expect.

STEP ONE: EXCITEMENT

You are planning your trip overseas and you feel excited about all the new people, places, and things you will experience. Upon arriving at your destination you probably feel ecstatic about living abroad. Everything is exhilarating for the first few weeks or months. Believe it or not, you are in the first phase of culture shock, where everything seems great, maybe even "better" than where you came from. This euphoric phase is fun and can last awhile—weeks, maybe even months.

STEP TWO: DEALING WITH YOUR ETHNOCENTRICITY

Then, gradually, you start making a mental list of all the little things that are getting on your nerves: it takes forever and a day to get anything done; the people are lazy, rude, or boring; the morning commuters are packed liked sardines in the trains; and the list goes on. It finally starts to sink in that acting the same way you do at home doesn't necessarily get you the same results you would get at home, results you have taken for granted all these years. When you are up against language barriers and cultural norms that you still don't fully grasp, depending on where you live, smiling at your neighbors, the butcher, and the post office clerk may earn you more frowns or unwanted attention than you bargained for.

My unconscious smiling mechanism, common to many Americans, always seemed to kick in at the (unbeknownst to me) most inappropriate moments. The corner butcher in Germany probably still thinks I wanted more than a few slices of turkey, all because of my inability to order with a straight face! And I thought by smiling I was just being friendly. Early one morning I attempted to politely greet a neighbor who lived in my apartment building. I was shocked and confused when he responded to my upbeat "Guten Tag" with a strange and bewildered look. I had seen him on numerous occasions in passing. Didn't he recognize me? Was it too early in the morning? Later my German boyfriend informed me that it is not common to greet your neighbors, especially to greet them heartily at 7:00 A.M. It would be seen as stepping over that imaginary line into my neighbor's private space. The poor guy probably spent the whole day trying to figure out who in the heck I was. After you make a few of these embarrassing and uncomfortable cultural and behavioral blunders, it is no wonder things start to get on your nerves.

Your ethnocentric attitude, which I'm sure you don't think you have, starts to appear. When we head overseas, we tend to overpack, that is, we tend to unwittingly schlep along a lot of emotional baggage. At first you might think the infrastructure of your home country is superior and don't understand why things in your host country don't run as smoothly as they do at home. At this point, some of us global adventurers start feeling disillusioned, confused, and frustrated. Why? Because the values we've internalized are being turned upside down and challenged by a culture we don't yet understand.

This happened to me and it happened to nearly all of my friends who have lived overseas. In Germany I always heard foreigners complain about the shops not being open twenty-four hours a day, as they are in the United States. In Portugal, foreigners complain about the lack of disciplined drivers, garbage in the streets, and persistent tardiness. These attitudes imply that one way is better than another. However, this kind of ethnocentric thinking simply reflects a way of life that you have grown accustomed to. Living abroad challenges you to adapt to alternative (not necessarily better or worse) ways of doing things. So, although the shops in Germany close early by U.S. standards and it takes some serious time-management skills to fit shopping into your day, it is because Germans value the separation of work and play. They have a "work to live" mentality. They tend to work seven- to eight-hour days so they have time to enjoy their *Feierabend,* or free evening, to relax, play sports, and spend time with family. Now *that* is a way of life many Americans wouldn't mind getting used to.

An ethnocentric attitude limits your experience of the world. Since you are a guest in foreign territory, you should do your best to not only learn the language, but to understand the mentality of those with whom you come into contact. Reading up on your new country's cultural behavior before you arrive can help you avoid embarrassing cultural faux pas and gain confidence more quickly.

STEP THREE: UNDERSTANDING STEREOTYPES

As your ethnocentric attitude escalates, you enter the most critical phase of culture shock. Ethnocentricity breeds an "us versus them" mentality that leads to stereotyping. As you know, stereotyping is a behavior generalization about a certain culture or ethnic group that is used to draw conclusions about that culture. Stereotypes can be positive, negative, or neutral, but in this phase, stereotyping tends to be negative. As your frustration with your host country mounts, your once elated emotions may swing to the other end of the pendulum, where you start to feel mild annoyance to pure dislike for the country, the government, or its people. You start to use sweeping generalizations to describe the country, its food, the mentality of the people, and so on.

If you go abroad you will undoubtedly be faced with stereotypes about your own nationality or the country you are visiting. The American stereotype, especially in northern Europe, is that we are loud, obnoxious, and fat; we eat too much McDonald's; and we have a poor ability to learn foreign languages. Of course, the grain of truth in every stereotype does not justify the exaggerated and ludicrous meanings that many people come to believe. In the same vein, you should not believe everything you hear about the French, the Germans, the Japanese, or any nationality or ethnic group.

Getting to know the locals personally will help you appreciate them for who they are and see the validity of their views of the world. This is a critical point. Unfortunately, some people abandon their international endeavors too early and hightail it back to their familiar home surroundings, bringing with them negative memories, unfair stereotypes, and a less-than-enlightening personal experience.

How do you deal with stereotyping? Keep an open mind to understand the full picture. If you don't speak the language fluently or understand the cultural values, you are not in a position to draw accurate conclusions. Accepting people the way they are can only enhance your adventure. Be yourself, stay flexible, and help educate others about your viewpoints and experiences so that they can understand *you* as a unique individual. People are people wherever you go in the world. Your training as a citizen of the world includes practicing humility, understanding, and compassion. If you notice yourself displaying ethnocentric behavior, using negative stereotypes to define your experience, or seriously questioning your ability to come to grips in your host country, it may help to realize that you are smack dab in the middle of culture shock. Stick it out a bit longer and you'll eventually move into the next phase.

STEP FOUR: ACCEPTING AND DEALING

After some time, perhaps after reading up on the culture, getting to know some locals, and becoming more proficient with the language, you will feel more comfortable in your surroundings. It can take awhile to really understand the cultural mentality behind the behaviors of people from different ethnic upbringings. For example, when you start to understand and accept that southern European cultures, while not as fast or efficient as North American societies, have a very strong family orientation, you may actually begin to appreciate or prefer elements of your new home. Indeed, after a longer stay in one country, the language, the culture, and its norms might start to feel like second nature. Congratulations! You have faced culture shock and still managed to have a successful experience abroad. Think that's all there is to it? Not quite. See what happens when you finally return home.

STEP FIVE: REVERSE CULTURE SHOCK

You can imagine what happens when it comes time to return to your real home. Leaving your second home can be just as traumatic as it was to leave your home country. Reverse culture shock happens when you have to readjust to your own country, its people, and the mentality. It used to feel quite natural, but now it may feel strange and foreign. It can take a few months to a few years to fully acclimate to your own country again.

Don't be surprised if you go through the phases of culture shock in your own country. You may feel ambivalent about reentering your society after experiencing an exciting life overseas. Difficulty readjusting can be compounded by your friends' and relatives' superficial or limited interest in listening to stories about your adventures, especially if they have never lived abroad. Your experiences abroad were educational; they gave you new knowledge about yourself. You aren't the same person you were before you left, so it's natural to feel different when you return.

The culture-shock/reverse-culture-shock phenomenon is experienced to some degree by nearly everyone who lives abroad for an extended period of time. To help you work through and reflect on your life abroad, it can be rewarding to share your experiences, pictures, and memories with other global citizens who also have exciting

A Global Citizen's Perspective

Reverse Culture Shock

by Elizabeth Kruempelmann (ekruempe@hotmail.com)

I experienced reverse culture shock when I returned to the States after a year studying abroad in Denmark and Germany. My senses were overloaded the whole time I was abroad. In addition to my business studies and field trips, I traveled to fifteen countries, experienced the fall of the Berlin Wall, learned two languages, and lived with a Danish family, Danish students, and a German family. My senses were stimulated almost every minute. When I returned home, there was simply nothing that stimulated my senses anymore. My mind had been programmed to speak foreign languages every day. Suddenly I had no outlet for all the words I was used to using regularly. My body had been accustomed to soaking up different sights and sounds, and suddenly I had to readjust to life on Maple Street. Except that Maple Street didn't seem like the Maple Street I remember.

I had grown unaccustomed to so many things that settling into the American rhythm took some time. I had to get used to the way Americans interact with each other, like how we talk to strangers while waiting in line at the checkout counter. I had to learn how to drive a car again, instead of taking public transportation. And I had to get used to eating at big chain restaurants, instead of quaint cafés and pubs. People who knew I had studied in Denmark would erroneously ask if I had learned Dutch. Or, mistaking Denmark for the capital of Sweden, would inquire about what the Swedes were like. And the whole time the only thing I could think of was how I would ever live abroad again. I belonged out in the world. Traveling and living overseas gave me the food that my mind needed to thrive and that I needed to be happy. I talked to everyone I could find who had lived abroad; we shared our hearts and souls and ideas for getting back to the life we all so desired.

Read more about Elizabeth Kruempelmann's international experiences elsewhere in this chapter as well as in chapters five, seven, and eight.

international tales to tell. You will eventually settle into your new routine at home, just as you settled into your life overseas. Not everyone will want to go back overseas immediately, like I did, but if you decide you want a taste of the global life again, then the ideas in this book will help you achieve that goal.

CULTURE-SHOCK RESOURCES

★ **Expat Expert**
www.expatexpert.com

This is a good Web site for finding out more about culture shock and its reverse effects. You will find several very good and funny articles, especially for the expat spouse.

★ **Culture Shock Series**
(Graphic Arts Center Publishing, recent editions)

This series provides helpful tips and proper etiquette for many destinations, including Australia, Borneo, India, Philippines, Spain, and Thailand. There are also a wife's guide and a parents' guide as part of the series.

★ **Homeward Bound: A Spouse's Guide to Repatriation**
by Robin Pascoe and Carlanna Herzog (Expatriate Press Limited, 2000)

Trying to resettle the kids and your partner after living overseas can be a challenge for many spouses. This book relates to the spousal repatriation challenge with tips and advice for the whole family.

★ **So, You're Coming Home**
by J. Stewart Black, Ph.D., and Hal B. Gregersen, Ph.D. (Global Business Partners, 1999)

Coming home after a life abroad is more difficult than most people realize. This book tells you what to expect and gives guidance for readjusting to your home country.

CHAPTER THREE

International Goal Planning: Survey Yourself

Let him that would move the world first move himself.

SOCRATES

There are lots of options to help you go where you want in the world. Do you spend $2,000 USD on a ten-day cultural trip to Machu Picchu, Peru; spend the same amount of money on a three-month backpacking tour in South America; or scout out opportunities to spend a longer amount of time working or volunteering abroad? Do you go back to school for an international MBA, or do you hop the next plane and find yourself a job once you get there? The self-survey section that follows will help you decide. Making your international experience fit with your personal, professional, and financial goals is what this chapter is about. In it you'll decide what kind of global experience would suit you best, taking into consideration the many options, your time frame, and your budget. Most people who succeed at having an international experience go back for more. They become global citizens and encourage others to do the same. The rewards are endless.

Survey Says!

This short self-survey will help you organize the information you found out about yourself in the last two chapters, pinpoint your goals, and plan for them wisely. There are no right or wrong answers. Use the information to clarify your desires.

INTERNATIONAL GOALS

These questions will help you decide where to start looking for international experience. If your main reason for going abroad is to gain knowledge for the sake of expanding your horizons, you may be interested in educational travel, taking courses overseas, or volunteering. If you want your international experience to help you get into a certain school or offer you more promising career prospects, then you might pursue study-abroad programs, internships, trainee programs, volunteering, or work. If you want to do something good for the world or make a difference for someone else, you might best achieve this by volunteering abroad. Regardless of the kind of global adventure you choose to pursue, you will undoubtedly have the opportunity to practice a foreign language, build a network of contacts, and have a lot of fun.

Pick your top three choices.

1. What is the main reason I want to go abroad?
   - ❑ To gain knowledge for knowledge's sake
   - ❑ To get into a good college or grad school
   - ❑ To be qualified for a certain career
   - ❑ To have better career choices or more job offers
   - ❑ To make more money or get promoted
   - ❑ To do something good for the world or to make a difference in someone's life
   - ❑ To have fun, stay active, and enjoy life
   - ❑ Other ______________

2. What do I expect to gain by going overseas?
   - ❑ Personal growth, a new perspective on life, or broader horizons

- ❑ Professional development, a certificate of study, global skills, or work experience
- ❑ Satisfaction from helping others
- ❑ An awareness of world issues
- ❑ Language skills
- ❑ International friends or professional contacts
- ❑ Other ____________
- ❑ All of the above

DESTINATION

Knowing where you want to go is the first step to getting there. Choosing a destination will help you narrow your Internet searches for travel itineraries, study or learning programs, internships, and volunteer opportunities.

Mark your target destinations and list the countries, cities, or specific places you hope to visit.

3. Where do I want to go?
   - ❑ Asia
   - ❑ Africa
   - ❑ Europe
   - ❑ North America
   - ❑ South and Central America
   - ❑ Australia
   - ❑ Antarctica/the Arctic
   - ❑ Other

Which countries?

Which cities or towns?

Specific place (like the Red Sea, Mount Kilimanjaro, the Alps, a Brazilian rain forest, an island in the South Pacific, etc.)?

TIME FRAME

Knowing how quickly you want to go abroad is important for prioritizing your travel options. Educational travel and some volunteer programs can be booked with just a few weeks' notice. However, if you want to apply for a study-abroad program, scholarship, fellowship, internship, or volunteer program like the Peace Corps, you'll need to apply three months to a year in advance. The same is true if you are looking for a job or trainee program that will send you overseas.

The amount of time you want to stay overseas may determine the avenues you pursue. If you want to stay longer than your travel, study, work, or volunteer stint allows, you might start thinking about how to combine several different options, such as studying abroad for one semester in one country, and then doing an internship the next semester in a different region. Or you might consider extending your time abroad to travel or look for a job.

Pick your top choice.

4. When would I like to go?
   - ❑ Less than 3 months
   - ❑ Within 6 months
   - ❑ Within 1 year
   - ❑ Within 1 to 3 years
   - ❑ Within 3 to 5 or more years

5. How long do I want to stay abroad?
   - ❑ 1 month or less
   - ❑ 1 to 6 months
   - ❑ 6 months to 1 year
   - ❑ 1 to 3 years
   - ❑ 3 to 5 years

- ❑ 5 years or more
- ❑ As long as possible
- ❑ Undecided

FINANCING

Money is an important factor to consider ahead of time. If you have to fund your adventure yourself, try to get scholarships or travel grants to pay for part of it before dipping into your own savings or applying for a loan. Also, make sure you take advantage of frequent-flier miles and travel discounts. If you just want to cover your expenses, there are internships and volunteer opportunities that pay room and board and may even offer a stipend. If you are volunteering abroad, you might try to implement some fund-raising ideas like organizing a raffle, selling merchandise, or asking for travel contributions from friends and family. Consider teaching, doing professional exchanges, or following your dream career overseas if you want to earn money by working abroad. See chapter nine for tips and advice on funding your travel and living expenses.

Choose your top three financing options.

6. How do I prefer to finance my international experience?
   - ❑ Savings
   - ❑ Loans
   - ❑ Discounts through frequent-flier miles and special offers
   - ❑ Scholarships, fellowships, or travel grants
   - ❑ Stipends or paid living expenses
   - ❑ Income from a job abroad
   - ❑ Money raised from sponsors (e.g., selling T-shirts and candy to raise funds)

YOUR STATUS

There are travel, study, volunteer, and work options for everyone. Students, retirees, and physically challenged people, in particular, are the lucky ones. You're eligible for a plethora of travel discounts, so be sure to always ask. Your professional status might also be relevant when applying for a work permit or arranging contract jobs.

Mark all that will apply to your status when you go abroad.

7. I will be a . . .
   - ❑ Student
   - ❑ Professional
   - ❑ Consultant, freelancer, or contractor
   - ❑ Work-at-home parent
   - ❑ Retiree
   - ❑ Physically challenged person

LIVING CONDITIONS

Knowing your living preferences will be important when choosing an educational trip or when planning to live abroad for a long period of time. If you prefer to go low budget when traveling, you could stay in local pensions or pop up your own tent in a campsite. If you're used to being pampered, a high-class luxury hotel might be more your style. Alternatively, you could be more immersed in the culture if you stay in renovated castles and historical places of interest. If you choose to study or intern abroad, you will have to decide ahead of time if you want to stay with a family, live with students, or find a place of your own. If you are one of the lucky ones and are sent abroad by your company to work or if you are participating in a paid internship or volunteer program, your housing might be paid for by the sponsoring company or organization.

Choose your top three living preferences.

8. Which type of living conditions do you prefer for your next experience abroad?
   - ❑ Luxury or comfort hotels
   - ❑ Bungalows or huts
   - ❑ Pensions
   - ❑ Student housing
   - ❑ Bed-and-breakfast inns
   - ❑ Your own apartment

- ❑ Homestay (staying in the home of a local family)
- ❑ Shared apartment
- ❑ Castles, convents, or historical places
- ❑ Camping
- ❑ Company-paid housing

YOUR INTERNATIONAL SKILL SET

Becoming a true global citizen means strengthening your global skills. If your long-term goal is to live a rich and meaningful life full of global adventures, then choosing an education, profession, and hobbies that develop your international skills is essential.

Check off the global characteristics you are still working on. (See pages 2–7 for definitions of these characteristics.)

9. Which of the following skills do you need to obtain or strengthen?
   - ❑ Curiosity about the world
   - ❑ Hobbies and interests with an international element
   - ❑ Awareness of global issues
   - ❑ An international network of contacts
   - ❑ Involvement in and contribution to global causes
   - ❑ Language proficiency
   - ❑ Cross-cultural appreciation and adaptation
   - ❑ A global way of thinking
   - ❑ International travel experience
   - ❑ An international education
   - ❑ International work experience
   - ❑ Sharing international experiences

YOUR PERSONALITY

Traveling and living abroad can be challenging. Understanding your personality will help you choose appropriate overseas experiences. A risk-oriented individual may enjoy active travel ventures and exotic cultures, like a multisport ecotrip in Nepal and Bhutan. A risk-averse, career-oriented individual may prefer experiences that are closer to home, like a prearranged career-related internship in an English-speaking country where the cultural differences aren't so extreme. Knowing your personal limits will help you make the most of your global sojourns.

Choose the answer that best describes your personality.

10. Which situation better describes you?

Comfort in new situations:

- ❑ I like meeting people anywhere, any time, in any language.
- ❑ I find it difficult or intimidating to meet people when I don't understand the language or culture.

Career versus fun:

- ❑ I am willing to be flexible about my work abroad as long as I'm having fun.
- ❑ It is most important to me to work hard and develop a progressive career path while I'm abroad.

Risk orientation:

- ❑ I would rather venture overseas on my own and have a flexible itinerary that I can design to fit my schedule.
- ❑ I would rather go abroad with a program that arranges all the logistics and itineraries.

Open-mindedness:

- ❑ I am generally open to new cultures, people, food, travel experiences, and opportunities.
- ❑ I am particular about where I travel, who I meet, what I eat, and the opportunities I pursue.

INTERESTS

The following themes will help you narrow down the substance of your next trip. They are grouped under three main categories: Culture, Environment, and World Causes. You may think of other areas of interest that you would like to make part of your travels, like scuba diving or native crafts. Read through travel brochures and descriptions of exchange programs to help you focus your interests.

Choose the five topics that sound the most interesting to you.

11. Which of the following themes do I want to explore during my next trip abroad?

Culture: People, Life, Business	**Environment: Nature and Outdoor Adventure**	**World Causes**
Archaeology, anthropology	Adventure travel	AIDS education
Cultural traditions	Conservation	Building homes for locals
Famous people	Deserts	Community service
Food, wine, beer	Ecotravel	Cooperation
Government	Endangered species	Democracy
Interaction with locals	Environmental awareness	Disease control
International business, commerce, law	Health	Education and training
Language and cross-cultural communication	Mountains	Grassroots development
Local art and history	National parks	Human rights
Museums	Natural wonders of the world	Peace
Music	Nature reserves	Political issues
Religious customs	Oceans	Population control
Social events	Outdoor sports	Poverty
Sports	Physical fitness	Religion
Technology	Rain forests	Social responsibility

Theater and the arts	Research expeditions	Sustainable earth
Wonders of the world	Spirituality	Women's issues
World capital cities	Wildlife	World health
Other	Other	Other

SKILLS

We all have a unique set of skills, characteristics, and experiences. The key to a successful global experience is knowing which of your many skills you want to use and develop. If you are searching for a way to work abroad, for example, you may not realize that your native English language abilities are a true skill that is in demand worldwide. But not everyone wants to use their English language skills. You may want to focus on your Spanish language skills or your computer programming experience. Deciding which skills you want to use or develop abroad will make planning your trip that much easier.

Being aware of what you have to offer has another advantage. Traveling overseas is not just a learning experience for you. Many times the people you meet will want to learn as much from you as you do from them. So, whether you plan to use your carpentry skill to build schools in Africa or just relate stories from home to the people you meet, you are offering something in return for the privilege of visiting a foreign culture. It is this type of cross-cultural *exchange* that makes for unique and memorable adventures abroad.

Circle all of the skills and characteristics you want to use abroad. In addition, list any other specific skills or experience you would like to leverage.

12. What skills, characteristics, or experience do I have to offer?

Insights on my culture	An open mind to learn	Language skills
Creative talent	Problem solving	Manual skills
Analytical thinking	Leadership	Project management
Spiritual insight	Friendship	Teaching ability
Cooperation	Sensitivity	Computer skills
Your specific talents:	Your specific experience:	Your specific skills:

CAREER FIELD

Your international career field is important for several reasons. First, it will help you narrow down study options, scholarships, professional exchanges, internship and training opportunities, volunteer programs, and professional jobs overseas. Second, it will give you a frame of reference for thinking about long-term career opportunities that could develop from your overseas work. For instance, if you are going abroad to teach in the Peace Corps, it is helpful to think about how the contacts you make could someday provide you with job connections or how Peace Corps volunteers are given preference for certain graduate study programs and jobs.

Choose the field in which you hope to study, intern, volunteer, or work abroad. (This may be different from your current field. For example, if you are a lawyer going overseas to teach, mark the "teaching" field.) Make a note if you specialize in a specific branch of your field. For example, if you are an investment banker, you would note this in the International Banking slot. As this list is not exhaustive, simply add your career field under Other, if your field isn't listed.

13. What is your international career field?

Career Fields	Notes
BUSINESS	
Corporate Business	
International Banking	
International Consulting	
Internet, E-Commerce	
Advertising	
Journalism	
Broadcasting	
Other	
FEDERAL GOVERNMENT	
USAID	
CIA (Central Intelligence Agency)	
DOD (Department of Defense)	
Peace Corps	
Other	
NONPROFIT	
Educational Exchange	
Environment	
Disaster Relief	
Other	

CAREER FIELDS	NOTES
INTERNATIONAL LAW	
TEACHING	
OTHER	

YOUR INTERNATIONAL ADVENTURE

As you've probably figured out by now, there are myriad ways to go abroad. Following is a list of some of the options described in this book, organized by four main categories: Educational Travel, Learning Overseas, Volunteering Overseas, and Working Abroad. Check off the areas that sound the most interesting for your next trip.

14. What type of travel most interests you for your next trip overseas?

Educational Travel

- ❑ Adventure travel
- ❑ Artistic workshops
- ❑ Cultural, environmental, and nature travel
- ❑ Homestays
- ❑ Language vacations
- ❑ Socially responsible travel
- ❑ Independent travel
- ❑ Special travel: Adults over Fifty and the Disabled

Learning Overseas

- ❑ Study abroad—traditional and direct enrollment programs
- ❑ International business (MBA/MIM) programs and executive education
- ❑ International scholarships, fellowships, and research grants
- ❑ International exchanges for teachers and educational administrators

- ❑ Exchanges and training abroad for professionals
- ❑ Foreign language training

Volunteering Abroad

- ❑ Archaeology
- ❑ Business and economic development
- ❑ Community development
- ❑ Cultural programs
- ❑ Environmental protection and research
- ❑ Health programs
- ❑ Humanitarian relief and service programs
- ❑ Teaching English
- ❑ Volunteering online

Working Abroad

- ❑ Internships and volunteering
- ❑ Trainee and international management trainee programs
- ❑ Teaching abroad
- ❑ Short-term work and au-pair programs
- ❑ Find a job with a company that will send you overseas
- ❑ Go overseas and get a job
- ❑ Create your own opportunity abroad
- ❑ Get an international job at home

This self-survey should have helped you narrow down your geographical preferences, time frame, financing options, career interests, skills, and options for going overseas. Whether your drive to get international experience is fueled by a passion for the global stock market or the romantic French language, you now have a basis for planning your future. I will discuss different programs available for going abroad in part II. The next five chapters might give you ideas you hadn't previously considered, so feel free to come back and retake the survey to see if your answers have changed.

Next . . . the final details to make it all happen. Read on!

PART TWO

A World of Opportunity

We are a product of the choices we make, not the circumstances that we face.

ROGER CRAWFORD

If you are like me, you have a burning passion to experience the world. But is it really that simple to just pick up and live in a foreign country? Are you wondering if you'll be able to find the right opportunity? Don't worry, this section of the book will answer those questions and squash all doubts. You *can* just pick up and leave (with some careful planning, of course), and there is a whole world of opportunities out there. This part of the book will give you a good idea of the wide range of options available—from traveling around the world to building an exciting global career.

If you already know that working overseas is your real dream, you're probably wondering how this book is going to help you get there. You may have even flipped anxiously to the working abroad chapter to get a quick peek at what's involved with getting a job abroad. At this stage you might not be interested in the other chapters on studying, volunteering, and traveling abroad. But wait . . . how are you going to work in South America if your Spanish is rusty at best? A culture and language trip may be just the experience you need to beef up your language skills. How will you find a job with an international development agency like USAID if you've never even been to a developing country? Gaining a few years of volunteer or Peace Corps experience abroad could help you get your foot in the door. How will you land a high-level international business job that will take you around the globe if all you have is a bachelor's degree in biochemistry? Getting an international business degree at an international business school like INSEAD could be your ticket. Finding a job overseas involves getting your international skills and experience up to speed through traveling to your destination, obtaining an international education, and getting initial work experience through internships and volunteering. So go ahead and turn to the work chapter, but be sure to read the other three, too!

The range of opportunities that fall within each chapter is more extensive than the chapter titles would lead you to believe. Educational travel, for instance, consists of everything from adventure tours in Africa, to homestays in Ireland, to expeditions in the Amazon.

Learning abroad ranges from traditional study-abroad exchange programs worldwide to scholarships and fellowships.

Volunteering abroad can mean joining traditional grassroots community service projects in developing countries, helping out on a marine research expedition, or working for a short time in a Mexican village as you travel around that region.

International work includes overseas internships and volunteer exchanges, being sent on an overseas assignment, or working in an international capacity in your home country. There are lots of ways to gain international experience.

Your job right now is to analyze each option, comparing the goals of the program to your personal goals. If you want to spend several weeks next summer learning about the Brazilian rain forest, there are various travel, study, and work options to consider. Do you take an educational trip that includes lectures and visits to villages? Do you arrange a course for credit at a Brazilian university? Do you do an internship on environmental and rain-forest issues? Do you volunteer in a Brazilian rain-forest community? The great thing is that it's possible to do any and all of the above. So check out the next five chapters to gain a clear understanding of all of your options.

THE QUICKEST WAY TO SEE THE WORLD

If you are anxious to go abroad and don't care what you're doing, as long as you are living or traveling in a foreign country, here are the quickest ways to get there:

- Pack your bags and just go . . . Bon Voyage! (See "Independent Travel" in chapter four and "Short-Term Work Abroad" in chapter seven.) Also, pick up a copy of *Work Your Way Around the World* from Vacation Works.
- Book an around-the-world flight with your backpack in hand. Get your ticket today from Airtreks.com! (See "Independent Travel" in chapter four.)
- Study a language overseas . . . and do some volunteer work while you're learning. (See "Language Vacations" in chapter four and "Cultural Programs" in chapter six.)
- Live, study, or work abroad through an exchange program. (See CIEE on page 127 and Studyabroad.com on page 129.)

THE CHEAPEST WAY TO SEE THE WORLD

If you want international experience, but don't want to finance it yourself, these are your best bets:

- Teach abroad. (See WorldTeach on page 212.)
- Apply to a fully paid scholarship program. (See Fulbright on page 156 and Rotary on page 162.)
- Do an internship or a traineeship that includes a stipend. (See CIEE on page 127, AIESEC on page 233, and AIPT on page 233.)
- Volunteer with a program that pays your expenses. (See the Peace Corps on page 209.)
- Work abroad. (See chapters seven and eight.)

CHAPTER FOUR

Educational Travel: Life-Seeing and Sight-Seeing

Don't let school interfere with your education.

MARK TWAIN

Have you ever traveled to a special place that made such a deep and lasting impression on you that you came home feeling invigorated, enlightened, and changed? That is exactly what educational travel is about—helping you see the world from a different and enriching point of view. Whether you backpack around Europe, take a wildlife safari in Kenya, or meet indigenous tribes in the Amazon, traveling to other countries transforms you. As opposed to pleasure travel, like a beach vacation in Florida, educational trips focus on personal and cultural exchanges that will stay with you for years to come. Educational travel is an experience outside your own country that increases your awareness of other cultures, languages, ways of thinking, and global issues.

A few years ago a cousin of mine who had never been abroad before mustered up the courage to do something a bit out of the ordinary. On a whim, she decided to visit a poor Mexican village with her church group. The plan was to stay with a Mexican family for five days and engage in a cultural exchange. She had no idea that these five days would have a profound and emotional impact on her.

Most of us don't consider what it means to live in poverty until we experience it. My cousin stayed in a two-room hut with six people, without electricity or hot water. What she sacrificed in comfort, however, was made up by the outpouring of generosity from her Mexican host family. The love and emotional bonding she felt with her host family and their children in those few days was like nothing she had experienced in the United States, despite coming from a very loving family. She was treated like a queen and was given the only bed in the house while the parents and four children slept on the floor. Refusing their overwhelming hospitality would have insulted them.

The most remarkable aspect about this short cultural exchange was that despite the language barrier—my cousin didn't speak Spanish and her host family didn't speak English—my cousin was forever changed by the gracious and unselfish love shown to her by this family.

Educational travel is a value-added approach to travel. Why "value-added"? Because educational travel focuses on learning something new and exciting or doing something challenging and different that adds value to your life. So instead of squeezing all the cultural stuff in during a marathon sight-seeing trip or mindlessly baking yourself on some foreign beach, try a multisport adventure, partake in a painting workshop, or learn a new language. Educational travel is for people of all ages, including families, retirees, and the disabled. On educational trips you can learn more about a subject, improve your skills in a specific area, or simply have a new life adventure. Sponsors of educational travel include colleges and universities, museums, church and community groups, environmental and voluntary service organizations, educational exchange organizations, and commercial tour companies.

A WORD ABOUT ECOTOURISM

Ecotourism is a buzzword in the travel industry today. Unfortunately, when many of us travel, we think more about having a good time than about the potentially negative impact we are having on the wonderful people and exotic places we visit. The environmental, cultural, and economic impact of tourism over the years has created a host of problems with far-reaching and long-lasting effects. Ecotourism strives to help communities around the world sustain themselves by supporting local goods and services without negatively impacting the natural and cultural areas being visited.

According to the International Ecotourism Society (TIES), the Kenyan safari industry, for example, has caused an alarming decrease in some animal populations, food supplies, and habitats. Tribespeople have been forced from their native lands to make room for the expansion of tourist lodges and safari tours. Mass tourism and large resorts in many famous tourist locations around the world consume natural resources and overburden the land, leading to blackouts, local water shortages, sewage problems, and pollution in waterways, among other problems.

What's being done to solve these global problems? Fortunately many tour operators are embracing low-impact, ecoconscious tourism. Worldwide efforts toward developing the ecotourism industry are paying off. Tribespeople of Africa are gaining more control over the use of their land. They operate ecolodges and serve as tour guides and experts to the local tourism industry, while conserving their traditions and homeland. Large hotel chains are starting to employ people from the community and are allowing local crafts to be sold in their tourist shops. Initiatives are being taken in Nepal to educate and train women to serve the trekking tourism industry.

Increasing awareness of the impact tourism makes on fragile resources will help us be more informed and responsible travelers. Thus, ecotourism means traveling responsibly, supporting local industry by staying at locally run lodges, buying from local shops, and taking care not to disturb the natural environment. Ecotour operators assist the cultures they visit by contributing a portion of their profits to preserving the community, culture, and environment. In a world where large, multinational companies tend to run smaller, family-owned shops out of business, it is vital to the survival of small villages and communities that ecoconscious travelers support the local economy. Ecotourism allows the adventure travelers of the world to pursue their global passions while having a positive influence on the cultures they visit. Many types of adventure travel companies offer low-impact ecotours. To learn more about ecotourism and ecotour companies, see the International Ecotourism Society's (TIES) site at www.ecotourism.org.

The Benefits of Educational Travel

The benefits of educational travel go beyond sight-seeing. Educational travel offers a wonderful opportunity to check out a city or country before you move there, pursue your hobbies overseas for a minimum investment of time and money, expand your worldly horizons, or learn a new skill.

GET TO KNOW A COUNTRY BEFORE YOU MOVE THERE

Whether your goal is to study, work, or retiree overseas, it is always a good idea to visit your dream destination before picking up and moving there. You might find the notion of living in a small French village lined with cobblestone streets and a thousand-year-old cathedral quite romantic, until you find out that the nearest movie theater and restaurant are fifty miles away. You might also discover that tucked away in that little French village is a three-hundred-year-old mansion you would like to buy, restore, and turn into a bed-and-breakfast. Traveling to your dream destination gives you the opportunity to scope things out ahead of time. Pay attention to the cost of living; real estate prices; proximity to supermarkets, doctors, restaurants, and entertainment; and the state of the local economy.

EXPLORE YOUR INTERESTS INEXPENSIVELY

Educational travel offers international exposure without long-term investments of time or money. If you only have a few weeks of vacation per year, you can still explore your international interests and hobbies inexpensively through homestays, house exchanges, or discount travel options. With the vast array of travel options available and the strong competition between travel agencies, travel overseas doesn't have to be expensive these days.

EXPAND YOUR HORIZONS

Learning more about a subject of interest can stimulate your imagination and expand your awareness of the world. Studying with experts in the field gives you the opportunity to ask questions and engage in enlightening on-site discussions. And there are educational travel opportunities to suit all interests. If you are curious about world religions, you could join a pilgrimage or tour to the Holy Land. If you are interested in archaeology or art history, you could accompany archacological experts on tours of Mayan ruins in Peru or visit art masterpieces in the famous churches of Italy with an art historian. Ecology, the environment, metaphysics, architecture—the possibilities are endless.

IMPROVE SPECIFIC SKILLS

Are you an aspiring chef? Do you want to finally beat your boss at golf? Is your dream to become skipper of your own yacht someday? Educational travel is a perfect way to develop a specific talent or skill. In addition to improving your cooking at a gourmet school in Italy, practicing your swing at the world's most famous golf courses, or getting your yachting license in the Caribbean, you can learn about corporate team building, fitness, hiking, horseback riding, languages, mountain biking, rafting, rock climbing, scuba diving, trekking, and more. That beats a class at your local community college any day, doesn't it?

How to Avoid the Pitfalls of Educational Travel

Educational travel can be a fun and exciting way to see the world. However, there are a few pitfalls you should be aware of that have the potential to turn a dream trip into a real nightmare.

SELECT A REPUTABLE ECOTRAVEL OPERATOR

Select tour operators who support low-impact, environmentally conscious travel that sustains local cultures and economies. You will not be sacrificing the quality of the tour, but rather you will be supporting the quality of life in the areas you visit. The International Ecotourism Society's (TIES) site at www.ecotourism.org provides a list of ecofriendly tour operators worldwide.

MAKE SURE YOUR LEADERS ARE EXPERTS

The last thing you want to find out is that you know more about the subject than the "expert" tour leader you have paid to teach you. You should be able to ask fairly in-depth questions and learn from your guide's knowledge and field experience. If you are planning your travel through a tour company, ask about the tour leader's training and qualifications, and find out if this person is a certified guide.

TAKE THE NECESSARY PRECAUTIONS

If you are traveling to a country with a turbulent political situation, check the U.S. Department of State's Web site (www.state.gov) to find out about travel warnings and what to do if you need help. Try to ascertain how the people in your destination area generally feel about travelers from your country and your government. Learn to be sensitive when talking about religion and politics.

Get the proper vaccinations far enough ahead of your departure. When my husband and I went on safari in Zimbabwe for our honeymoon, we had to get several rounds of vaccinations. Despite discovering this over two months before our trip, we still just barely had time to fit in all the shots.

Be aware that if you are traveling to certain areas of the world, especially developing countries, during certain times of the year, the chances of contracting diseases might be high. Although we traveled to Zimbabwe just before the malaria season, we still stocked up on Deet mosquito spray as an extra precaution against malaria. Better safe than sorry!

Planning Your Educational Adventure

With most types of educational travel, you plan your trip the same way you would plan a vacation. Contact your travel agent to arrange a flight, book your accommodations, and arrange the program details. However, it may be easier to book the trip directly through the organization offering the program. Luckily, most of the programs can be booked online.

OK, now that you know *how,* let's talk about *what.* Ready to find out what kinds of programs are out there?

Getting Started: Using the Internet to Research

If you want a general overview of the different types of travel options available, I have found the best place to start is an "umbrella" Web site that has a comprehensive database of educational tour operators and programs. My favorite sites are Shaw Guides, the Specialty Travel Index, and Transitions Abroad. They all have extensive databases of unique educational trips worldwide. You can read about hundreds of educational travel organizations and the programs they offer from a quick search of the listings.

★ **GoAbroad/Adventure Travel Abroad**

www.goabroad.com or www.adventuretravelabroad.com

Adventure Travel Abroad, an affiliate of GoAbroad.com, makes it easy to find the best travel adventures. Search by country or type of trip (climbing, hiking, rain forest tour). The site also lists links to the State Department's travel warnings, travel guides, embassies, currency converters, travel insurance companies, and accommodations.

★ **Shaw Guides, Inc.**

www.shawguides.com

P.O. Box 231295
Ansonia Station
New York, NY 10023
Phone: 212-799-6464
Fax: 212-724-9287

Shaw Guides features an online database of more than forty-three hundred sponsors of learning vacations and creative career programs throughout the world, including arts and crafts workshops, cultural travel, cooking schools for professionals and hobbyists, golf schools and camps, photography, film and new media schools and workshops, language vacations, writers' conferences and workshops, tennis schools and camps, and high-performance workshops. You simply pick your area of interest and then view the listings either by country or by the month in which you'll be traveling.

★ Specialty Travel Index

www.specialtytravel.com

305 San Anselmo Avenue, Suite 313
San Anselmo, CA 94960
Toll-free: 800-442-4922
Phone: 415-459-4900
Fax: 415-459-4974

The Specialty Travel Index is the number-one source for adventure and special interest vacations worldwide. You can search by geographic destination, by tour operator (with over six hundred listed), or by activity (such as anthropology, hiking, volcano tours, yoga programs). The A–Z activity index is extensive and can connect you with specific agencies that offer the types of tours you want in your preferred destination. The activity index includes links to nearly every type of educational travel mentioned in this book: adventure, cultural, educational, disabled, eco, family, homestay, expedition, senior, student, study, and language programs.

Transitions Abroad Publishing
www.transitionsabroad.com

P.O. Box 1300
Amherst, MA 01004
Toll-free: 800-293-0373
Phone: 413-256-3414
Fax: 413-256-0373

Transitions Abroad contains travel program listings by country or region for teens, adults, seniors, and the disabled, as well as sites for immersion and responsible travel. You can subscribe to *Transitions Abroad* magazine, which lists individual travel, and study and work programs overseas.

Your Educational Travel Options

An educational trip abroad can ignite your passions, stimulate your mind, and transform your attitudes. Let your passions and personal goals guide you to the best travel choice. There are thousands of ways you can get an education while traveling. Here are some of them.

ADVENTURE TRAVEL

Would you rather visit a world-famous zoo or go on safari to view elephants and rhinos up close? Would you enjoy a leisurely bike ride along a coastal boardwalk, or does your heart race only when you think about hiking, biking, and backpacking in the Andes? If you answered the latter in both cases, you are an adventurer at heart. Traveling in adventure mode is just your style. To find an exciting adventure trip to your dream destination, start by getting an overview of the many types of adventure programs available by searching Adventure Quest or GORP Travel. Or check out the more specialized tours also listed below.

GET SCUBA CERTIFIED

Through the Professional Association of Diving Instructors (PADI) or the National Association of Underwater Instructors (NAUI), you can become a licensed scuba diver, take specialized courses like underwater photography and shipwreck diving, or become a Certified Dive Master and teach at certified dive schools around the world.

PADI (Professional Association of Diving Instructors) Worldwide

www.padi.com

30151 Tomas Street
Rancho Santa Margarita, CA 92688
Toll-free: 800-729-7234
Phone: 949-858-7234
Fax: 949-858-7264

PADI offers the following recreational and technical dive courses:

- PADI Experience Programs
- PADI Scuba Diver
- PADI Open Water Diver
- PADI Adventure Diver
- PADI Advanced Open Water
- PADI Medic First Aid
- PADI Rescue Diver
- PADI Master Scuba Diver (Rescue and Master Divers are the pros!)

Contact PADI for information on Project AWARE, PADI educational programs, and about becoming a certified scuba diver.

NAUI (National Association of Underwater Instructors)

www.nauiww.org

P.O. Box 89789
Tampa, FL 33689

Toll-free: 800-553-6284
Phone: 813-628-6284
Fax: 813-628-8253

NAUI lists scuba-diving jobs worldwide on their Web site, along with course information and international locations. The following is NAUI's Progression of Training Courses:

- NAUI Skin Diver
- NAUI Scuba Diver
- NAUI Advanced Scuba Diver
- NAUI Master Scuba Diver
- NAUI Specialty Diver Courses
- NAUI Technical Diver Courses
- NAUI Leadership Courses
- NAUI Scuba Instructor

★ Adventure Quest

www.adventurequest.com

482 Congress Street, Suite 101
Portland, ME 04101
Toll-free: 800-643-5630
Phone: 207-871-1684
Fax: 207-871-1684

Adventure Quest offers fourteen hundred adventure tours, including animal treks, cultural tours, nature viewing, language learning, cooking schools, houseboat voyages, and more. Special tours are available for families, gays and lesbians, seniors, singles, and the disabled. You can search tours by destination, type of tour, or special tour categories.

★ **Arctic Odysseys**

www.arcticodysseys.com

2000 McGilvra Blvd., East
Seattle, WA 98112
Toll-free: 800-574-3021
Phone: 206-325-1977
Fax: 206-726-8488

Do you crave something original . . . an adventurous travel experience that only a few people ever will take advantage of? Arctic Odysseys provides high-quality travel experiences in the Arctic regions of Canada, Greenland, Siberia, and China as well as in Antarctica. The rolling tundras and sparkling glaciers of the Arctic beckon to wildlife enthusiasts, expert skiers, dogsledders, and others drawn to the breathtaking beauty. You may stay in a comfortable hotel or a comfortable igloo, depending on your trip. Groups are small and special interest groups are welcome.

Backroads

www.backroads.com

801 Cedar Street
Berkeley, CA 94710
Toll-free: 800-462-2842
Phone: 510-527-1555
Fax: 510-527-1444

If you want to travel in style, Backroads has over 142 trips to 101 destinations worldwide. Their trips focus on biking, walking, cross-country skiing, golf, multisport vacations, and even cooking classes. With Backroads, you can still have adventure without "roughing it." Backroads specializes in five-star-quality adventure trips at world-class resorts.

GORP TRAVEL

www.gorptravel.gorp.com

P.O. Box 1486
Boulder, CO 80306
Phone: 877-400-GORP
Fax: 303-444-3999

For active and adventurous travel, search GORP Travel's activity index, which includes multisport vacations, adventure cruises, wildlife tours, and more. You can also search for adventures by destination or special interest tours, like women-only packages, 50-and-better tours, family vacations, or solo adventures.

HIMALAYAN TRAVEL, INC.

www.gorp.com/himtravel.htm

8 Berkshire Place
Danbury, CT 06810
Toll-free: 800-225-2380
Phone: 203-743-2349
Fax: 203-797-8077

Himalayan Travel offers adventure travel to not only to Nepal and Tibet, but also to India, Pakistan, Bhutan, Southeast Asia and Indochina, South America, Africa, and the Middle East. You can choose to stay in comfortable hotels or partake in a more strenuous expedition.

ARTISTIC WORKSHOPS

If your creativity needs a jump start, there are many different art programs and workshops in inspiring places from Venice to Guatemala. Whether your artistic talents involve filming documentaries, creating pottery, painting landscapes, writing novels, cooking, or another form of artistic expression, you can develop your creative side with one of the following programs.

★ Apicius Institute

www.apicius.it

Via Guelfa 85
50129 Florence
Italy
Phone: (39) 055-287-1-43
Fax: (39) 055-239-89-20

The Apicius Institute in Italy offers professional programs in hospitality management, professional wine expertise, and the culinary arts. Each program includes internships in local restaurants, wineries, or hotels.

★ Art Workshops in Guatemala

www.artguat.org

4758 Lyndale Avenue, South
Minneapolis, MN 55409
Phone: 612-825-0747
Fax: 612-825-6637

Workshops for artists and wanna-be artists are offered on the following subjects: ceramics and pottery, decorative arts, drawing, basketry, bookmaking, folk art, garden design, painting, photography, film, video, precious gems, sculpture, and wood. Photography and video classes include adventure travel photography, fine art photography, documentary photojournalism, and landscape photography. Writers can take classes in creative writing, fiction, and poetry. You can even learn the ancient art of carpet making. In the small colonial town of Antigua, Guatemala, experienced teachers and artists conduct twenty-five art programs at all levels for groups of up to ten people. Costs range from $1,725 to $1,925 USD, and include airfare from most major U.S. cities, along with family stays, breakfasts, and field trips.

★ Booktowns of Europe

www.booktownwriters.com

P.O. Box 1626
West Chester, PA 19380
Phone: 610-486-6687
Fax: 610-486-0204

"Booktowns" are small, rural villages that generate 80 percent of their income from book-related industries, including rare-book dealers, book fairs, auctions, and used-book stores. Booktowns of Europe may be snuggled away in a thirteenth-century village in the Pyrenees, a medieval town on the Dutch-German border, or an eighteenth-century glacier town on the Norwegian fjords—the ideal setting to nurture your writer's spirit for one to two weeks. In writers' workshops in France, Germany, Holland, and Norway, you will learn how to organize your research; outline materials; develop the background, characterizations, and plot; and discover the tricks of good writing. In groups of twelve people or less, you spend two hours in class and the rest of the time either seeing the local sights or completing assignments related to the workshop. A one-week workshop that includes ground transportation, lodging with breakfast and dinner, and one excursion costs $1,650 USD. Discounts are available for seniors, teachers, full-time students, and groups.

★ Cuisine International

www.cuisineinternational.com

P.O. Box 25228
Dallas, TX 75225
Phone: 214-373-1161
Fax: 214-373-1162

Cuisine International offers culinary experiences in Italy, France, Portugal, England, Brazil, the United States, and other locations worldwide.

Weeklong cooking classes provide outstanding regional experiences in art, history, and culture. Many of the classes are held in private homes or small, unique hotels. The Web site includes information about the renowned chefs who teach here, as well as details about each region and the program costs.

International Art Landscape-Painting Workshops

www.ARTw.org

Tjasa Demsar Painting Workshops
Palazzo Ottolenghi
San Marco 2569
30124 Venice
Italy
Phone: (39) 041 241 101 2
Fax: (39) 02 700 426 753

International Art Landscape-Painting Workshops are designed for beginning hobbyists as well as aspiring professionals. You'll learn landscape-painting techniques from an internationally acclaimed artist, get art history lessons, and take painting excursions to inspiring museums, castles, gardens, and small towns. Tours take place in Italy, Mauritius, Slovenia, and France. There are fifteen one-week programs throughout the year. Costs range from $500 to $1,950 USD, and include lodging, excursions, materials, and half of your board.

La Meridiana

www.pietro.net

Pietro Elia Maddalena
Bagnano 135
50052 Certaldo (Florence)
Italy
Phone: (39) 0571660084
Fax: (39) 0571660821

La Meridiana, set among the vineyards and olive trees in Tuscany, offers a perfect source of artistic inspiration. Internationally recognized ceramic artists lead one- and two-week ceramic, pottery, and sculpture seminars and workshops for beginners to advanced levels. Workshop courses range from throwing to sculpture. The first week is spent learning how to use the wheel and other special techniques, and the second week is spent finishing, glazing, and firing. Six classes are offered between June and September. Visits to Florence, Rome, and Venice can be arranged. The two-week program costs about $1,100 USD and includes materials, lodging at a Tuscan farmhouse, and lunches on class days.

Le Cordon Bleu, Inc.

www.cordonbleu.net/index.htm

404 Airport Executive Park
Nanuet, NY 10954
Phone: 845-426-7400, Ext. 130
Fax: 845-426-0104

Le Cordon Bleu is a world-famous culinary school that was originally founded over one hundred years ago in Paris. Now it has ten worldwide locations where you can take a variety of culinary classes to suit your taste. Earn a bachelor's degree in restaurant management or cooking certifications such as the Cordon Bleu Diploma or Grand Diplome. Scholarships may be available. Check the Web site for contact information of a school near you.

Siesta Tours South, Inc.

www.siestatours.com

P.O. Box 90361
Gainesville, FL 32607
Toll-free: 800-679-2746
Fax: 352-371-8368

Siesta Tours specializes in travel to the colonial town of Oaxaca, Mexico. In addition to visits to artisans' workshops, tours include trips to the UNESCO World Heritage Sites in historic Oaxaca and the pre-Columbian archaeological site of Monte Alban. Siesta Tours can organize workshops for artists, photographers, and travel writers; a continuing education program for teachers; or a Spanish immersion class.

CULTURAL, ENVIRONMENTAL, AND NATURE TRAVEL

Many tour operators offer a variety of educational trips that combine cultural education, environmental awareness, and nature travel. On cultural tours you will visit famous monuments and landmarks, sample a region's culinary delicacies, and meet with the local people, be they indigenous tribes or village schoolchildren. In many parts of the world, like the Black Forest in Germany or the Himalayan mountains in Nepal, the natural environment plays an important role in the culture. In these cases, cultural trips can easily turn into nature and environmental adventures that might include activities like bird-watching in the country or a wildlife safari in the jungle.

★ Far Horizons Archaeological & Cultural Trips

www.farhorizon.com

P.O. Box 91900
Albuquerque, NM 87199
Toll-free: 800-552-4575
Phone: 505-343-9400
Fax: 505-343-8076

Explore the ancient Mayan cities of the Yucatán Peninsula, visit the magnificent temple complex at Angkor Wat in Cambodia, or marvel at the classical world of Sicily's archaeological treasures. Far Horizons participants meet privately with renowned experts and local inhabitants in each destination, and dine on the local delicacies, drink traditional beverages, learn a few words of the language, and travel to areas rarely visited by tourists. Hotels are frequently restored historical buildings reflecting the charm of the country. Far Horizons donates to scientific projects and indigenous community programs, thus creating friendships and partnerships from which you as a traveler will benefit.

★ THE INTERNATIONAL ECOTOURISM SOCIETY (TIES)

www.ecotourism.org

P.O. Box 668
Burlington, VT 05402
Phone: 802-651-9818
Fax: 802-651-9819

TIES, a membership organization for ecotourism professionals, has resources to help raise awareness about ecotourism, connect travelers with TIES-member tour operators and ecolodges, and encourage potential members to join ecotravel organizations and networks.

★ INTERNATIONAL EXPEDITIONS, INC.

www.ietravel.com

One Environs Park
Helena, AL 35080
Toll-free: 800-633-4734
Fax: 205-428-1714

International Expeditions, an ecotourism and conservation organization, is a world leader in nature travel, including expedition voyages. Maybe you've seen something like this on the Discovery Channel: remote islands where penguins, sea lions, and seals congregate in their native habitats. This is one of the many wonders you might experience on the "Epic Voyage of Discovery" tour in Patagonia and Tierra del Fuego. International Expeditions sets the standards for environmentally sensitive travel by promoting conservation, environmental awareness, and research through their expeditions and programs.

Journeys International, Inc.

www.journeys-intl.com

107 Aprill Drive, Suite 3
Ann Arbor, MI 48103
Toll-free: 800-255-8735
Phone: 734-655-407
Fax: 734-655-2945

Journeys International is an ecotourism agency that offers worldwide exploratory, family, and specialty travel, including nature and culture explorations to forty-five countries. Imagine exploring caves and lava fields, sampling edible rain-forest plants, and sliding down rocks into beautiful waterfall pools on the Samoa Treasure Island Odyssey tour, geared especially toward families. Journeys International also offers expedition cruises to Arctic regions. Trips are led by local, English-speaking guides, who are often botanists or ecologists, and who have extensive cultural knowledge, as well as an awareness of the current social and political events.

Myths and Mountains

www.mythsandmountains.com

976 Tee Court
Incline Village, NV 89451
Toll-free: 800-670-6984
Phone: 775-832-5454
Fax: 775-832-4454

Myths and Mountains designs group itineraries based on five educational themes: cultures and crafts, religion and holy sites, wildlife and the environment, learning journeys, and folk medicine and traditional healing. Tour Tibet or Bhutan with a Buddhist monk, wander the ruins of Peru's Machu Picchu, or boat in the bays of Vietnam. Myths and Mountains has a collaborative program with CARE (see page 358), in which CARE donors and travelers learn how these types of international nongovernmental organizations operate abroad.

★ **NatureQuest**

www.naturequesttours.com

398 West Colorado Avenue, Suite GW
P.O. Box 22000, PMB 128
Telluride, CO 81435
Toll-free: 800-369-3033
Phone: 970-728-6743
Fax: 970-728-7081

Every NatureQuest expedition of ten to fourteen travelers is led by a staff of professional naturalists and is designed to create minimal impact on the environment. The focus is on the natural history, culture, and wildlife of each destination. NatureQuest's activities include sea kayaking, canoeing, hiking, trekking, wilderness camping, expedition cruises, white-water rafting, horseback riding, wildlife encounters, whale watching, mountain biking, and safaris. There are five levels rating physical activity and accommodation, from an easy-walking, no-camping tour to a fast-paced, strenuous outdoor camping adventure. For example, a fairly easy "Jungle Lodge and Riverboat Safari" tour is offered in Brazil or a fairly strenuous "Hiking and Paddling" expedition is offered in Chile's Patagonia region. Search tours by geographical destination or activity on their Web site.

HOMESTAYS

Simply put, a homestay is an opportunity to stay in a local family's home. But many homestays offer more than just a bed. In fact, the idea behind homestays is to give both the host family and the traveler a chance to interact. It's your opportunity to get an inside, personal, and unique introduction to a country and culture. Your host family may show you the local sites, offer you a typical home-cooked meal, engage in political discussions, or answer the many questions you might have about their country, history, and traditions. Homestays might also be part of language, study, or work programs.

When I took part in an intensive six-week German class in Braunschweig, Germany, I stayed with a lovely German family. As their daughter was married to an American and lived in the United States, they were happy to open their home to an American to learn more about American culture as well as to share their German traditions. My host family invited me to meet their German friends, showed me around their historic town, and shared typical German meals with me. And living with a family was the perfect way for me to practice my language skills too.

★ American International Homestays (AIH)

www.spectravel.com

P.O. Box 1754
Nederland, CO 80466
Toll-free: 800-876-2048
Phone: 303-642-3088
Fax: 303-642-3365

Enjoy local hospitality, food, and friendship when you stay with an AIH host family. They will not only act as your guide and interpreter, but will include you in their circle of family and friends. You can expect your English-speaking hosts to provide a private room, a home-cooked breakfast and dinner, and a guided tour. AIH can also arrange customized and group trips, transportation, guided tours, translators, and language study.

★ Cultural Homestay International (CHI)

www.chinet.org

2455 Bennett Valley Road, Suite 210 B
Santa Rosa, CA 95404
Toll-free: 800-395-2726
Phone: 707-579-1813
Fax: 707-523-3704

CHI arranges outbound group homestays for people ages thirteen and up. CHI also offers academic-year and semester study-abroad programs, career exploration "job shadowing," total immersion language classes, and au-pair stays abroad.

★ **United States SERVAS, Inc.**

www.servas.org

11 John Street, Suite 407
New York, NY 10038
Phone: 212-267-0252
Fax: 212-267-0292

SERVAS is an international, nongovernmental organization dedicated to understanding, tolerance, and world peace. The SERVAS program offers travelers the opportunity to meet with people of other cultures and to share in their daily life. Hosts can offer to accommodate someone for two nights or serve as a day host and guide. Participants must be at least eighteen years old and have an introductory interview, after which they receive a letter of introduction to show their hosts upon arrival. Interested travelers receive a list of SERVAS hosts they can contact in their destination. SERVAS has an open-door policy of providing hospitality to approved travelers from every race, creed, and country.

LANGUAGE VACATIONS

If you want to practice your Spanish, Russian, or Dutch, and you don't have enough time to take part in a long-term study-abroad program, you can easily immerse yourself in a short-term program through a language vacation. Language vacations are a fun way to soak up history and culture, get to know the locals, and practice an important life skill. As there are thousands of language schools worldwide, it is important to pick one with a good reputation for small-group immersion courses taught by qualified teachers.

A Global Citizen's Perspective

Spanish-Language Vacations

by Christina Picardi, K–5 ESL Teacher (Travelme32@aol.com)

As a preschool teacher in a bilingual classroom, I use my Spanish-language skills on a daily basis. Because they weren't as strong as I'd like, I decided to spend a couple of my summers off studying Spanish abroad. The first summer I signed up for a program through Centro Linguistico Conversa (see page 102), an organization that arranges Spanish-immersion classes in Costa Rica.

I spent a month living with a family in a suburb of San Jose, and attended semiprivate (two students per one professor) classes for five hours a day. I spent the evenings with my "Tican" family. My Costa Rican "mama" was a foster mom, so there were always three to four babies or toddlers living with us. She was a caregiver for babies awaiting adoption, mainly to other countries like the United States and Canada. She was a very warm and loving woman, who had one daughter of her own. I enjoyed spending my evenings with them after a long day at school.

Since I was only there for one month, on the weekends I tried to see as much of the country as possible. One weekend, some friends and I went to Playa de Manuel Antonio, a beach on the west coast, and another weekend we traveled to a mountainous area called Monteverde, where most of the rain and cloud forests are. I also spent a memorable evening in the hot springs of the Arenal volcano.

I found myself more fluent in Spanish and better able to keep up with the parents of my bilingual students after studying in Costa Rica. So, three summers later I decided to head for Quito, Ecuador. I loved Costa Rica and had a very positive experience there, but I wanted to try another country and another family placement. Through an acquaintance, I learned of an organization called Language Link (see page 104), which sponsors language learning in Central America, South America, and Spain. This time, I wanted to do more than sit in a classroom, so I chose their more adventurous ANACONDA program. I lived on an island, five hours from Quito, in a hut without electricity or hot water. I stayed there for six days, learning Spanish in the mornings and taking day trips around the island in the afternoons. We visited a zoological society and an herb farm, and I spent some time in Quito with a family sponsored by the school.

Both of these experiences abroad were invaluable, introducing me to hands-on living and learning in two very different Latino cultures. I have more confidence in my work as a bilingual teacher and enjoy better relationships with my students. There are so many great options for studying abroad; I can't wait for next summer.

Read more about Christina Picardi's international experiences in chapter five.

Berlitz Study Abroad

www.berlitz.com

400 Alexander Park
Princeton, NJ 08540
Toll-free: 800-457-7458
Fax: 609-750-3593

Berlitz, one of the most reknowned language schools in the world, offers summer study-abroad language-immersion programs in English, French, German, Italian, Japanese, and Spanish. Dates and prices vary per program. In addition to language study abroad, Berlitz also offers online language learning for selected languages, and cross-cultural courses to learn the proper dos and don'ts of living in a foreign culture.

Centro Linguistico Conversa

www.centralamerica.com/cr/school/conversa.htm

Apdo. No. 17, Centro Colon 1007
San José, Costa Rica
Toll-free: 800-354-5036
Phone: 506-221-7649
Fax: 506-233-2418

At Conversa, students are placed in small groups of four or less and learn Spanish for up to five hours per day. Students can either stay with a host family, in an on-campus lodge, or in an on-campus family suite. Child care and children's programs are available. Conversa is ideally located in the hills, with fresh air, fruit trees, and a breathtaking view. There are many sports activities, like swimming, hiking, and volleyball. Conversa's Super Intensive Spanish studies have been approved by the College Consortium for International Studies (CCIS), which means students can earn six or seven credits for a four-week cycle. Costs range from $650 USD for one week to $2,500 USD for a four-week program.

★ China Advocates

www.chinaadvocates.com

1635 Irving Street
San Francisco, CA 94122
Toll-free: 800-333-6474
Fax: 415-753-0412

China Advocates arranges five- to forty-week lessons in Mandarin Chinese taught in Beijing. Classes are twenty hours per week and the program includes field trips to famous historical and cultural sites, and weekend excursions to nearby cities. Instructors are specially trained at teaching Mandarin to foreigners. For a five-week course, the cost is approximately $3,010 USD, including airfare, lodging in a shared dormitory, the language course, and cultural events.

★ École des Trois Ponts

http://3ponts.edu

Château de Matel
42 300 Roanne
France
Phone: (33) (0) 4 77 71 53 00
Fax: (33) (0) 4 77 70 80 01

École des Trois Ponts offers a unique opportunity to live in the eighteenth-century Château de Matel while taking French-language or French-cooking courses. Classes run from one to twelve weeks for six hours per day and are taught by French teachers with master's degrees in teaching French as a foreign language. Supplemental activities include a French country-cooking course, a wine-tasting class, swimming, tennis, horseback riding, and cultural visits to medieval villages near Roanne. Prices range from $800 to $1,450 USD per week, which includes lodging and meals.

Language Link, Inc.

www.langlink.com

P.O. Box 3006
Peoria, IL 61612
Toll-free: 800-552-2051
Phone: 309-692-2961
Fax: 309-692-2926

Language Link, a program of the Better Business Bureau, is dedicated to offering high-quality, complete-immersion Spanish language programs in Mexico, Peru, Ecuador, Costa Rica, Guatemala, and Spain, in addition to special courses for teens and executives. Language Link requires a high level of professional excellence from the language schools with whom they partner and the families they choose to host foreigners. They are dedicated to personal and affordable service. Language classes last four to seven hours per day, and include some excursions. Your participation in some programs will create jobs for the indigenous people and preserve the rain forests. Program costs will vary. Activities could include jungle walks, cooking classes, visiting places of interest, and sports. You may be able to receive course credit for this program, so be sure to ask if this is of interest to you.

Lingua Service Worldwide

www.linguaserviceworldwide.com

75 Prospect Street, Suite 4
Huntington, NY 11743
Toll-free: 800-394-5327
Phone: 631-424-0777
Fax: 631-271-3441

Lingua Service Worldwide is an independent agency focusing on full-immersion, intensive language vacations for one to thirty-six weeks year-round. In addition to sixty regular language programs, their special programs include full language immersion by living and learning at your teacher's home, summer camps for children and teenagers, discounted language training for flight attendants and air personnel, and intensive Spanish programs for the fifty-plus crowd. Most programs also include activities like a welcoming party, cultural tours and excursions, cooking and dancing classes, and cultural enrichment. Lingua's services include registering you for your course of choice, supplying course and accommodation information, providing free medical insurance for programs less than four weeks in duration, and taking care of your travel arrangements, if you wish.

SOCIALLY RESPONSIBLE TRAVEL

Are you concerned about promoting peace, human rights, and social justice? Do you want to see firsthand what is being done to promote sustainability in societies around the world? Traveling with an awareness of the social problems of the regions you visit and a desire to learn more about peace, social inequality, women's and children's issues, education, and health is what socially responsible travel is all about. Socially responsible travel is for people who desire a deeper understanding of these issues and others. You might visit with community leaders, scientists, or students to learn about the particular problems they are facing, or you may be able to intern or volunteer your time to help a local cause.

★ Cross Cultural Solutions

www.crossculturalsolutions.org

47 Potter Avenue
New Rochelle, NY 10801
Toll-free: 800-380-4777
Phone: 914-632-0022
Fax: 914-632-8494

In its Field Insight Programs, Cross Cultural Solutions offers a unique look inside a culture through meetings with tribal chiefs, social pioneers, traditional healers, and women's groups, or by visiting shantytowns, places of worship, and schools. Groups of fifteen people or less have access to rare encounters to which the average traveler would never be exposed. On the Insight India tour, groups meet the children at Mother Teresa's Home, stay at an Indian palace as the guests of an Indian prince, and have enlightening discussions with social workers in shantytowns. Cross Cultural Solutions also offers volunteer work opportunities to provide humanitarian assistance in China, Ghana, Russia, India, and Peru.

★ **Education Travel Seminar to Two-Thirds World**

Center for Global Education at Augsburg College

www.augsburg.edu/global

2211 Riverside Avenue
Minneapolis, MN 55454
Toll-free: 800-299-8889
Phone: 612-330-1159
Fax: 612-330-1695

The Center for Global Education is recognized for its work in experiential, intercultural, and educational travel opportunities to Mexico, Central America, the Caribbean, and southern Africa. Each year the center coordinates more than twenty short-term educational travel programs for adults over eighteen who are interested in issues of peace, justice, democracy, local and regional politics, human rights, and indigenous issues. Groups have less than twenty people. The goal of the program is to bring participants face to face with local and global conditions that promote or hinder peace and social justice, and with people working for a more just and sustainable world. The most unique dimension of the tours is hearing the perspectives of those working at the grassroots level for social change.

Travel seminars will show you the realities of life for most of the world's people, allow you to examine the root causes of poverty and injustice, transform your interpretation of the world, and help you become an agent for positive change. The style of a travel seminar is intensive—stimulating for the seasoned traveler, yet accessible to those who have never experienced life in Two-Thirds World communities.

You might visit a group of peasant farmers in the countryside or residents of a refugee camp, meet with government officials, or hear the views of the opposition political party. You might visit a school and talk to students and teachers about the education system, or visit with doctors or community health-care workers at a local clinic. Participants are housed two or more to a room in retreat centers, guest houses, or modest hotels. Seminars last one to three weeks and are offered year-round. Costs start at $595 USD and do not include airfare. Ask about travel grants.

★ **GLOBAL EXCHANGE/REALITY TOURS**
www.globalexchange.org/tours

2017 Mission Street, #303
San Francisco, CA 94110
Toll-free: 800-497-1994
Phone: 415-255-7296
Fax: 415-255-7296

Global Exchange is a nonprofit research and educational organization focused on economic, political, environmental, and social justice worldwide. Reality tours are socially responsible and educational study trips, fact-finding missions, activist delegations, and custom trips for academic, business, and professional groups. Reality tour groups serve as human rights delegations. Groups observe and report on events in areas of conflict and may act as election monitors. Tours include visits with community leaders, government officials, journalists, scientists, academics, doctors, writers, artists, musicians, dancers, environmentalists, religious leaders, and human rights activists. During their visits to various towns and cities each Global Exchange group holds discussions on the latest political developments, resource sustainability, gender issues, human rights, and the impact of economic globalization. On the Web site you can search for tours by issue, such as health and healing or civil rights, or by date. Tour destinations include Argentina, Cuba, Haiti, India, Iran, Ireland, Mexico, Palestine, Israel, and South Africa. Costs vary per program.

INDEPENDENT TRAVEL

There are lots of ways to travel independently. Many adventurous travelers design their own budget travel "hosteling" tours, homestays, house exchanges, and around-the-world trips.

How do you design a sojourn to fit your specific needs? It depends on where you want to go, how much time you have, and the mode of transport you prefer. When planning your own trip abroad, the vast possibilities for arranging flights, accommodations, and sight-seeing agendas at a fraction of the cost of a package tour are worth considering.

Getting to Your Destination(s)

Once you know where you want to go, you have to decide how to get there. If your destination is just one country or city, you can find special fares and offers at airlines, travel agencies, and through online promotions at travel-related Web sites. If you've got to work within a limited budget, try naming your ticket price on auction Web sites like hotwire.com or priceline.com. Alternatively, you can use frequent-flier miles to get a free or discounted ticket. (See "A Global Citizen's Best Bets for Travel Discounts" and "Ten Easy Ways to Rack Up Frequent-Flier Miles" on pages 322 and 323, respectively.)

If your adventurous travel soul wants to literally travel around the world, Airtreks.com's around-the-world special ticket prices makes planning flights to multiple destinations easy.

★ **Airtreks.com**

www.airtreks.com

442 Post Street
Suite 400
San Francisco, CA 94102

Toll-free: 800-350-0612
Phone: 415-912-5600
Fax: 415-912-5606

On Airtreks.com you can create your own customized trip to multiple continents or take advantage of special offers to Circle-Pacific, Circle-Atlantic, or Circle-Asia, all for a little more than the price of just one international ticket. Special airfares can be as low as $1,395 USD to visit roughly seven major cities on several continents.

Finding Accommodations

As you plan your travel itinerary, you'll also have to think about where you'll stay, how much it will cost, and if you need to book in advance. Once you get to your destination, you can stay in youth hostels (which, by the way, are for people of any age), pensions, people's homes, campgrounds, hotels, or even a friend's place.

★ Hostelling International

www.iyhf.org or www.hiayh.org

733 15th Street NW, Suite 840
Washington, DC 20005
Phone: 202-783-6161
Fax: 202-783-6171

Hostelling International (or Youth Hostelling International) is a network of forty-five hundred hostels in over sixty countries. Hostels have rooms for individuals, families, and groups. Staying at youth hostels when you travel ("hosteling") is a fun and inexpensive way to meet people and have a great time. The catch is you must be a member. Join the Youth Hostel Association in your home country and your membership card will be valid for all local and international hostels. If your home country is not a full member of IYHF or if you've lived less than twelve months in your country of residency, you can still become a full member by buying six "Hostelling International

Welcome Stamps" at the hostels where you stay. With six stamps your Hostelling International membership makes you eligible for discounts off coach fares, car rentals, currency exchanges, adventure tours, meals, books, Internet use, and many other things. You can book rooms in hostels worldwide up to six months in advance by calling them or booking on Hostelling International's Web site under "Reservations."

If you prefer accommodations that will serve as a base for your travels, consider a house exchange. House exchanges are a wonderful and relatively inexpensive way to spend a long amount of time in one region, while having the flexibility to design your own daily travel agenda. By living in someone else's house while you allow someone to stay in yours, you will have a home base to return to after a day, several days, or even several weeks of sight-seeing around the region.

★ **HomeLink USA**

www.homelink.org or www.swapnow.com

P.O. Box 47747
Tampa, FL 33647
Toll-free: 800-638-3841
Fax: 813-975-9825

HomeLink International is the world's largest home-exchange organization with over fifty years of matching home-exchange partners around the world. HomeLink has the largest home-exchange database and greatest distribution list worldwide. Each participating country in HomeLink International has a local representative. A Web-only membership, which includes photos of houses, costs $50 USD per year. Full membership, which includes three full-color directories, is $106 USD per year.

If you want your specially designed travel agenda to include cultural immersion so you can have an interesting cross-cultural exchange with the locals or so you can practice your language skills, arrange a homestay with a family. See page 97 for a list of organizations and more details.

If you prefer the comforts of a hotel during your travels, book in advance using a local travel agent or online booking service, such as expedia.com or orbitz.com. These services also allow you to arrange a rental car, if you need one. If you don't book in advance, you can always wing it and find a hotel upon arrival in your city. Just be sure to avoid peak times when hotel prices are at their highest and vacancies at their lowest.

Traveling Around

Independent travelers tend to be adventurous types, preferring to plan their days as they go. Buy a good city or country guide, like *Let's Go Europe,* Rick Steve's *Back Door* travel books, or an *Eyewitness Guide,* all of which can be found at your local bookstore or Amazon.com. A good guide will give you all the essential information you need for traveling: addresses and phone numbers of accommodations, restaurants and points of interest, opening times and costs of the main sights, advice for using public transportation, maps, travel tips, and more. You may also want to visit the local tourist office to get train and bus schedules and free maps. Some tourist offices can book accommodations and local tours for you too.

Whether you'll be spending a few weeks or a few months traveling on your own, you might also consider taking an intensive language class (see page 99), volunteering (see page 173), or doing some of the other things you'll read about in this book.

Whatever your reasons for traveling, there are at least two basic resources every globetrotter should never leave home without: a travel guide and a travel journal. Here are a few of my favorites:

★ **Culinary Travel Journal**
(Ten Speed Press, 2000)

If enjoying delicious local delicacies is one of the main reasons you travel, then this journal is for you. With sections on cooking schools, European restaurants, farmer's markets, wine regions, and more, you'll be able to record not only your travel adventures, but your culinary experiences as well.

★ Cybercafés: A Worldwide Guide for Travelers

3rd edition by cyberkath@traveltales.com (Ten Speed Press, 1999)

This is a handy little guide, especially if you'll be country hopping. It lists 750 locations worldwide where cybercafés allow you to surf the Net, check your email, and send a virtual postcard.

★ Eyewitness Travel Guides

(DK Publishing, various editions)

Eyewitness Travel Guides are my personal favorite, especially if you are going to live abroad for a long period of time. The color photographs make your destination come alive. Detailed information on the different regions of the country, the history, politics, traditions, travel tips, travel itineraries, food and drink recommendations, and contact information for accommodations and eateries is provided. You might pay a bit more for this travel guide, but it has it all. *Eyewitness* also has travel planners, language courses, phrase books with cassettes, and travel maps.

SPECIAL TRAVEL: ADULTS OVER FIFTY AND THE DISABLED

There are two groups of travelers that are unique because educational travel programs have been specially designed for their needs. People over fifty and disabled folks used to be at a disadvantage when it came to traveling overseas. Not anymore. Tourism is big business around the world. Accommodating the needs of all travelers is becoming an increasing trend in tourist centers worldwide. Now there are several very good organizations that arrange overseas study, travel, and volunteering opportunities for everyone who desires to see the world.

Adults over Fifty

The pace of many adventure tours and activities are not geared to the needs of the fifty-plus crowd, but the following organizations are.

★ ELDERHOSTEL

www.elderhostel.org

11 Avenue de Lafayette
Boston, MA 02111
Phone: 877-426-8056
Fax: 877-426-2166

Elderhostel offers short-term educational travel opportunities, specifically designed for adults age fifty-five and older. With Elderhostel, you will have access to programs at universities, national parks, museums, conference centers, and other learning sites in the United States, Canada, and ninety other countries. During the one- to four-week trip, you engage in lectures, discussions, and field trips, and attend liberal arts classes focused on stimulating the intellect. There are no prerequisites for the courses. This is learning for learning's sake, therefore, no tests or assignments are required either. The diverse menu of programs covers subjects from Gershwin to Bach, archaeology to meteorology, and filmmaking to bread baking. Participants can study in a classroom setting, a laboratory, aboard a ship, in the mountains, or a small town. The educational courses and discussions are thought-provoking and geared to older adults with a passion for learning and adventure.

★ ELDERTREKS

www.eldertreks.com

597 Markham Street
Toronto, Ontario M6G 2L7
Canada
Toll-free: 800-741-7956
Phone: 416-588-5000
Fax: 416-588-9839

Eldertreks is the world's first adventure travel company for people over fifty. The goal of the program is to promote genuine, noncommercial encounters with local people and nature's wonders. Locations include thirty-one destinations worldwide.

If you enjoy walking and outdoor activities, are curious and interested in learning about new environments and cultures, like to go off the beaten path, enjoy traveling with and meeting other people, and savor travel to areas where physical and social conditions are different, and sometimes very different, from conditions in North America, then Eldertreks is for you. The types of trips available include wildlife/birds, mountains/volcanoes, tribal-ethnic cultures, swimming/snorkeling, adventure, historical sites/temples, train travel, boat travel, unique transportation (elephants, camels, rickshaws, tramways, etc.), and exploratory trips. In Indonesia, stay with the rain-forest dwellers, the Mentawai; learn the monkey dance; and taste sago, a local drink. In Tibet, drive over the Karo La Pass at 15,419 feet (4,700 meters) and gaze down upon the turquoise waters of Yamdrok Tso. In Guatemala, climb the steps of Tikal to greet the morning sun.

All of Eldertreks' trips involve some walking, in groups of fifteen people or less. You can choose a trip with activity ratings from easy to challenging, but you should be in fairly good shape. Accommodation levels are mostly mid-range hotels and high-end guest houses and inns—all charming, comfortable, and safe. Restaurants range from small, local eateries to elegant retreats. Trips last eight to twenty-four days and are offered year-round. Costs range from $1,450 to $4,990 USD not including transportation to the destination. And if you would like to treat someone to an Eldertreks trip, gift certificates are available.

★ **Walking the World**

www.gorp.com/walkingtheworld

P.O. Box 1186
Ft. Collins, CO 80522

Toll-free: 800-340-9255
Fax: 970-498-9100

Walking the World is an outdoor adventure program with travel to over twenty-five destinations, designed exclusively for adults over fifty. The goal of the program is to provide a quality year-round schedule of reasonably priced, outdoor adventures specifically geared to the needs of older adults. Group size ranges from twelve to eighteen hikers, and is generally 60 percent singles and 40 percent couples. Hikes take place in national parks, rain forests, mountains, islands, and deserts, and take time to explore each area's plant and animal life, geology, weather, and human cultures. The "Italy: The Best of Sicily and Mount Etna" tour has twelve days of moderate hiking (six to nine miles daily), while staying in farmhouses and hotels. Hikers explore walking routes in the mountainous interior, follow coastal paths beside the Mediterranean in the Aeolian Islands, and even try some adventurous hikes on the volcanic slopes of Mt. Etna. Tours last seven to twenty-one days and are offered year-round. Costs range from $995 to $3,095 USD, not including transportation to the destination. Gift certificates are also available. To learn more, sign up for the free *Walking the World* online newsletter.

Disabled Travelers

In the past, tourist sites in many cities overseas might not have been equipped with wheelchair ramps, making it inconvenient for wheelchair users to get around. Fortunately, this is beginning to change. The following resources will help travelers of all abilities see the world.

★ Mobility International USA (MIUSA)

www.miusa.org

P.O. Box 10767
Eugene, OR 97440
Phone: 541-343-1284
Fax: 541-343-6812

MIUSA is a U.S.-based nonprofit organization with the mission to empower people with disabilities around the world through international exchange, information, technical assistance, and training. International exchanges for youths and adults concentrate on leadership training, community service, cross-cultural experiential learning, and advocacy for the rights and inclusion of persons with disabilities.

Training seminars, workshops, adaptive recreational activities, cross-cultural communication, language classes, and volunteer programs are available in Azerbaijan, Bulgaria, China, Costa Rica, East Asia, the United Kingdom, Germany, Italy, Japan, Mexico, Russia, and other countries. Both students and professionals with disabilities may also teach, study, research, work, and volunteer abroad. See the Web site to search a database of disability organizations worldwide, to get tips on financing your trip overseas, and to read travel stories from other MIUSA members. For professionals working in the disability fields of education, employment, rehabilitation, health, and community integration, there are professional exchange programs to interact with your counterparts worldwide. MIUSA also manages the National Clearinghouse on Disability and Exchange (NCDE), which educates people with disabilities and related organizations about opportunities for international exchange.

★ National Clearinghouse on Disability and Exchange (NCDE)

www.miusa.org

c/o MIUSA
P.O. Box 10767
Eugene, OR 97440
Phone: 541-343-1284
Fax: 541-343-6812

For anyone who has ever thought about teaching, studying, or volunteering overseas, or who has questions about how to make programs more accessible to people with disabilities, the Clearinghouse is here to assist you.

New Horizons

www.faa.gov/acr/dat.htm

Federal Aviation Administration
800 Independence Avenue, SW
Washington, DC 20591

In 1990 the U.S. Department of Transportation published regulations as part of the Air Carrier Access Act that ensure people with disabilities will not be discriminated against in the airline industry. These rules are intended to minimize the special challenges disabled people face when they travel. The New Horizons Web site offers information for air travelers with disabilities, including information on planning your trip, getting on and off the plane, and compliance procedures.

CHAPTER FIVE

Learning Abroad: Engaging Your Mind

Education is our passport to the future, for tomorrow belongs to the people who prepare for it today.

MALCOLM X

When I set out to write this chapter, I called it "Study Abroad." However, studying abroad is commonly thought to be an option for only high school, college, and graduate students. What I discovered while researching learning opportunities overseas is that there are programs for everyone, including professionals, retirees, and disabled people. And I don't just mean study-abroad programs. You can learn about your field of choice through fellowships, scholarships, research grants, professional exchanges, or special skills certification courses.

Gaining an international perspective on your academic or professional field by learning about it overseas may give you the credentials and qualifications necessary for future international work and career opportunities. Unlike educational travel, learning abroad is a more formal approach to education. In addition to learning something relevant about your specialty, you'll receive formal recognition, such as a diploma or certificate.

Whether you are a student spending a semester studying international business in Europe, a scholar taking a year to research Italian immigration problems as part of a Fulbright fellowship, or an environmental technologist on a professional exchange program in China, by learning overseas you will study your specialty in-depth and from a unique foreign perspective—all while getting an exciting jump start on your global career.

The Benefits of Learning Abroad

Studying abroad or learning more about your profession overseas obviously has many advantages. However, most people can't imagine beforehand how learning abroad will change their views, relationships, and careers. Here are just a few benefits you can expect.

CREATE NEW OPPORTUNITIES FOR STUDY AND WORK

Understanding your field from another angle will deepen your awareness of related areas, opening doors to professional opportunities that you may not have known about or considered while studying in your own country. For instance, if you study abroad in a general humanities program and all of your courses are in French, you may become quite proficient in the language. Instead of applying to law school, you might decide instead to become a certified French translator or go on to study languages.

Learning more about your profession in an overseas environment allows you to offer your employer a different perspective on creative problem solving; improves your language ability, cross-cultural communication skills, and professional or industry contacts; and provides you with a unique international experience to boot.

CHANGE YOUR POINT OF VIEW

Whether you are a high school student, an undergraduate, graduate, or professional, the subjects you are studying will be presented by students and professors who have different perspectives than you have. It's one thing to learn about European business at an American university taught by American professors, and it is quite another thing to study it in a European capital, where you are taught by prominent experts who can arrange company visits and field studies in other European and Eastern European cities.

Sharing your cultural perspective will no doubt lead to lively and enriching intellectual exchanges. When the Berlin Wall came down in 1989 and Eastern Europe opened its doors to the West, it was a historical turning point for Europe. During my study-abroad exchange in Denmark, my professors—many of whom experienced firsthand World War II, the building of the Wall, and the subsequent division of Europe into East and West—gave us their personal perspectives. My professors also explained how the downfall of communism would affect individual European countries from a more global economic, social, and political point of view. I never would have gotten this kind of insight in the United States.

Access to professors may also be different than what you are used to at your home university. My Danish professors occasionally joined the students after class at the local pub, and one professor even invited the entire class to his house for tea. Being able to chat with the professors in a casual manner about topics both related and unrelated to our course was so educational. We discussed a variety of topics affecting Europe at the time, from the fall of the Berlin Wall to the economic and political issues facing the European Union. The learning opportunities that exist outside the classroom can sometimes be even more rewarding than those in the classroom.

APPRECIATE DIFFERENT CULTURES

During a learning-abroad program you will interact on a daily basis with your host family, other students, teachers, and a local community that may deal with daily matters much differently than you. You'll notice similarities between your host culture and your own country, like the love for children and family that almost every culture shares, and you'll notice cultural differences, like the

Japanese sense of collective decision making and consensus, which is different from the American management style. Discovering the mentality that lies behind your new foreign friends' behaviors and staying flexible in new situations will make your adventure a success. You will learn to embrace alternative ways of thinking and evaluate your own culture's mentality and values more objectively, overcoming stereotypes and ethnocentric ways of thinking.

MAKE INTERNATIONAL FRIENDS AND CONTACTS

Regardless of the reason you go abroad, it is often the people you meet who make the experience truly special. Because learning overseas can be challenging at first (especially if it is your first time in a foreign country), friendships become quite meaningful. You will form special bonds with the people you meet during your time abroad, because this group is an essential support network. The fun, adventure, and hard times you experience with your new circle of international friends will give you memories to treasure for years to come. It is worth staying in contact with these friends since they will form the backbone of your international network of contacts that you will call on in the future.

How to Avoid the Pitfalls of Learning Abroad

Learning abroad is a wonderful opportunity to do something different in life. By avoiding a few of the common pitfalls, you'll maximize the time you spend overseas.

HANG OUT WITH THE LOCALS

Avoid the tendency to hang out only with people of your own nationality who speak your language. This inevitably happens when groups of people from the same country find themselves together in a foreign place—as was the case during my study-abroad program in Denmark. We were all Americans, and most of us were living abroad for the first time. We fumbled around, trying to make sense of our new cultural experience, and found that going through the highs and lows together somehow made it more bearable.

However, if you want to get to know the locals and practice the language, it is imperative that you break away from the comfortable group of your compatriots.

Seek out new friendships with the natives. I decided to do this during my second semester of study in Denmark. I moved into a Danish student dormitory, which gave me a vastly different view on Danish life. I went out to parties with my Danish friends and learned to appreciate the famous Carlsberg and Tuborg beers. And (usually over a few beers), we discussed everything from their open attitudes toward sex and drugs to environmental awareness and social equality. Making the effort to limit my contact with people from the United States was a necessary step to making friends with the Danes. If you take part in an educational program abroad, you can also meet local people through your host family, teachers, work colleagues, in student housing, and at community and church events. So get out there and mingle!

FIND A GOOD LIVING SITUATION

Avoid staying in a living situation that makes you uncomfortable or limits your freedom. Your family or roommate should allow you to get involved with activities and practice speaking the language, and have the flexibility you need to enjoy a successful experience in that country. Several students in my study-abroad program didn't enjoy living with the families with whom they were placed. Sometimes there was a personality or culture clash, and in other instances the family didn't involve the student in its daily life. The students who didn't switch to another host family ended up getting much less out of their study-abroad experience and may even have gone home with unpleasant memories of their time abroad. Fortunately, an incompatible family match can usually be prevented if you recognize early on that your living situation is not optimal. Inform your program coordinator and ask to be moved to a family that is a better match.

LEARN ABOUT THE PROGRAM BEFORE YOU GO

Avoid going abroad without first educating yourself about the study program, the country, and the culture. Make sure the academic level of the courses meets your needs. Some people view studying abroad as a break from their normally rigorous studies. And that's exactly what some students want in a study-abroad program—a chance to party, travel, and have fun. I was looking for a chance to study international business at a reputable overseas university. I compared DIS, Denmark's

International Study Program with the other study-abroad programs my university offered in international business. The DIS program was by far the most demanding and is very reputable throughout Europe. The program far exceeded my expectations and is one of the most interesting life experiences I've ever had. Just be sure you know what you're getting into.

If possible, talk to former program participants or people who have lived in that country. While speaking to a friend who had studied in Denmark, I learned a few important cultural differences. She explained how the Danes are much more open about nudity than Americans. I knew about topless beaches in Europe, so this wasn't a big surprise. But then she explained that if you actually wear a bathing suit at all on any given beach in Denmark, you'll immediately stand out as a foreigner! Well now, that's a cultural tidbit I didn't find in any travel guide.

Planning Your Learning Adventure

How do you arrange a study-abroad, international exchange, fellowship, or skills training program? First, define your reasons for going overseas to learn. Once you know why you want to learn abroad, it will be easier to narrow down your options. Do you want to:

- ❑ obtain academic credit or professional experience;
- ❑ receive an international degree or certification;
- ❑ study a certain subject or in a particular country;
- ❑ learn a new language and experience a foreign culture;
- ❑ or travel and just have fun?

OBTAIN ACADEMIC CREDIT OR PROFESSIONAL EXPERIENCE

This is often the most practical reason students choose to study abroad. You receive college credit toward your degree and you get international experience that looks good on a graduate school application or résumé. Universities, graduate schools, and employers increasingly select candidates who show interest and initiative in rounding out their academic studies with study or work experience outside of their home countries. If you are a working professional, doing a fellowship or work exchange in your field may be the best way to expand your skills.

RECEIVE AN INTERNATIONAL DEGREE OR CERTIFICATION

Having a diploma or certificate that proves one has studied and excelled at a certain discipline gives many people a great sense of accomplishment. Indeed, a certificate or diploma from overseas implies that you grasp your specialty from a global perspective, adding credibility to your accomplishments and supplying a unique way to stand out from your peers.

Fellowships for research and study are available through many universities at the master's and doctorate levels. Certifications offered by professional development programs organized in connection with well-known international institutions are highly valued in the workplace and attest to your qualifications as you continue to build your career.

STUDY A CERTAIN SUBJECT OR IN A PARTICULAR COUNTRY

Some subjects are best studied in specific locations. For example, Italian art is best observed in the impressive cathedrals, monuments, museums, and architectural ruins for which Italy is famous. Archaeology is best studied at archaeological sites in Egypt, Mexico, Peru, Europe, Asia, and elsewhere. European business can best be learned in the European capitals, such as Brussels, Berlin, and London. Chinese language is best practiced in China. As you research overseas study programs, evaluate where you can learn the most about your specialty.

LEARN A NEW LANGUAGE AND EXPERIENCE A FOREIGN CULTURE

Many cultures are made up of multiple ethnic groups with many languages and diverse traditions. If you come from a country that is less diverse in its linguistic and ethnic composition, immersing yourself in a foreign language and culture will give you a broader understanding of the world. Learning programs for language and cultural immersion exist in nearly every country on the planet. Generally, you will have the option of living with a family or with students, or in community housing, all of which will intensify your cultural assimilation. A multicultural background will give you fresh perspectives on life, and language fluency could open new doors for you professionally.

TRAVEL AND JUST HAVE FUN

If your adventure learning overseas is not too rigorous, you will have time to go sight-seeing, hang out at the pubs, and meet people from all over the world. That is a learning experience by itself.

Now you should have a better sense of what your priorities for learning overseas are. You're halfway there. In the next section, I'll discuss all the learning options.

Getting Started: Using the Internet to Research

Whether your learning experience overseas is for academic, professional, or personal reasons, the easiest way to get started is to decide what you want to learn about and where you want to study. Is there a good study-abroad or master's program in engineering in New Zealand? Are overseas research grants and fellowships offered to learn about your specialty in France? Is there a way you can interact with other people in your profession worldwide?

Answer these questions and get an overview of the types of programs available by searching one of the following umbrella sites. The Institute of International Education (IIE) is a good site to research study-abroad, fellowship, seminar, and professional-exchange options worldwide. The Council on International Educational Exchange (CIEE) is an exchange organization for high school, college, and graduate students; families; communities; or professionals. Studyabroad.com, part of the GoAbroad.com network of sites, makes it easy to search country- and subject-specific study-abroad programs, including graduate programs.

Many of the organizations listed in this section offer internships and volunteer and work programs as well as student exchanges and other services. It is worth checking out the Web sites to see the full range of services that a particular program offers.

★ All Abroad

www.allabroad.com

You can search All Abroad's extensive database to find study and work programs overseas or drop into their Lounge to get advice or swap stories with other travelers. You can also prepare for your trip using their Guidebook and even buy the merchandise you'll need for your travels.

★ Alliances Abroad

www.alliancesabroad.com

409 Deep Eddy Avenue
Austin, TX 78703
Toll-free: 888-622-7623
Fax: 512-457-8132

Similar to CIEE and Studyabroad.com, Alliances Abroad arranges study-abroad programs, language learning, internships, volunteering, and work abroad for people of all ages.

★ American Institute for Foreign Study (AIFS)

www.aifs.org

River Plaza
9 W. Broad Street
Stamford, CT 06902
Toll-free: 800-727-2437

AIFS is a cultural-exchange organization for high school and college students, faculty, and advisors. AIFS arranges au-pair work, study abroad, and educational trips.

★ Council on International Educational Exchange (CIEE)/Council Exchanges

www.ciee.org or www.councilexchanges.org/us

205 E. 42nd Street
New York, NY 10017
Toll-free: 888-268-6245
Phone: 212-822-2600
Fax: 212-822-2699

Council, or CIEE, is one of the largest international exchange organizations in the world with extensive program offerings: travel, study-abroad, and language programs; internships; summer jobs; volunteer and work abroad; homestays; programs for teachers; and more. CIEE is divided into three independent operating entities: Council-International Study Programs, Council Exchanges, and Council Travel, and is dedicated to help students and recent grads develop skills for living in a globally interdependent and culturally diverse world.

★ **GoAbroad.com**
www.goabroad.com

8 East First Avenue, Suite 102
Denver, CO 80203
Phone: 720-570-1702
Fax: 720-570-1703

GoAbroad is one of the best online search directories, with thousands of links to international academic programs, study-abroad exchanges, international universities, and worldwide language programs (see Studyabroad.com below). You begin your search by choosing a country or subject area. The site has important information about why you should study abroad, how to find the best program, and how to prepare for your experience. GoAbroad also has subsite links to information about internships (www.internabroad.com), volunteering (www.volunteerabroad.com), work (www.jobsabroad.com), adventure programs (www.adventuretravelabroad.com), languages, and teaching.

★ Studyabroad.com

www.studyabroaddirectory.com

GoAbroad's affiliate site, Studyabroad.com, has thousands of listings of study-abroad programs in more than one hundred countries. You can search by country, academic semester, year, or experiential, high school, or language programs. You can also find information on English as a Second Language and the Teaching English as a Foreign Language certificate programs, along with other services, forums, and products.

★ Institute of International Education (IIE)
IIE Passport/Living and Learning Abroad

www.iiepassport.org or www.iie.org/svcs/sartoc/htm

809 United Nations Plaza
New York, NY 10017-3580
Phone: 212-984-5400

IIE, a nonprofit organization, is the largest and most experienced U.S. higher education exchange agency. IIE Passport, an information service of IIE, is designed to quickly match your study-abroad interests with U.S. education abroad programs. You can search the extensive database of over five thousand organizations by entering your country of choice, subject of study, or more than thirty other criteria. You can research organizations and financial-aid options, apply for a passport, and access a wealth of other learning-abroad resources. Listings include academic-year or short-term study-abroad programs that accept students, graduates, and teachers as well as human resources professionals, economists, artists, journalists, scientists, and other professionals.

IIE administers Fulbright fellowships for U.S. and international students and scholars and over 250 programs on behalf of sponsors including the U.S. Department of State, the U.S. Agency for International Development, foundations, corporations, government agencies, international organizations, and development assistance agencies in the United States and abroad. Projects are available in areas such as business, trade, communication, education, health, languages, teaching, science and technology, educational testing, and corporate recruitment tools.

★ Transitions Abroad Magazine

www.transitionsabroad.com

Transitions Abroad Publishing
P.O. Box 1300
Amherst, MA 01004-1300
Toll-free: 800-293-0373
Phone: 413-256-3414
Fax: 413-256-0373

The *Transitions Abroad* magazine is a bimonthly guide to learning, living, and working overseas. It can be ordered direct through their Web site or you may find it in some bookstores and newsstands.

If you'd prefer to work through a book that lists thousands of study-abroad programs worldwide or if you need a reference guide, these are the most complete sources.

★ International Study Telecom Directory

(Worldwide Classroom, 1999)

This directory contains contact information for thousands of study-abroad programs, as well as many related Web sites, an international currency conversion chart, and travel information.

★ **Peterson's Study Abroad 2002**
(Study Abroad, 2002)

In addition to the extensive study-abroad program listings, it also includes information on receiving college credit, paying for programs, internships, and volunteering abroad.

★ **Vacation Study Abroad 2000/2001: The Complete Guide to Summer and Short-Term Study**
edited by Sara J. Steen (Institute of International Education, 1999)

A publication of the Institute of International Education, this resource contains over twenty-two hundred programs for students, professionals, retirees, and adult learners to study, research, and volunteer abroad.

Your Learning Abroad Options

There are lots of options for learning overseas, from studying the culinary arts in Italy or getting an MBA in Singapore, to observing the problems your professional peers face in the human services field in India. Learning abroad and exchange programs are not limited to students. If you are a professional looking for ways to learn more about your industry, consider international fellowships, research grants, executive education programs, professional exchanges, and conferences. Below you will find some of the most popular options for studying, taking seminars, and skills training overseas.

STUDY ABROAD—TRADITIONAL AND DIRECT ENROLLMENT PROGRAMS

Studying abroad as part of a high school, college, or university exchange program is by far the most popular means of learning overseas. Taking courses in a certain area of specialization, studying a language, and pursuing a degree at a foreign university are extremely rewarding experiences, and it is common nowadays for students from all majors to spend at least one semester studying overseas. Programs are offered in almost all fields of study, usually for a period of several weeks up to several years. High school students often take a year off to travel or study abroad before college, and university students generally study abroad during their junior or senior years.

Study-abroad programs target academic studies, and students receive credit, a certificate, or a degree for their participation. If you are a college student, find out about study-abroad options by taking a trip to your on-campus study-abroad office. The staff should be able to tell you which study-abroad programs are offered through your university.

If you are still in high school or you don't find a program that suits your needs through your university, you should be able to take part in another university's study-abroad program or an exchange program through an organization like the Council on International Educational Exchange (CIEE), and then have the academic credits transferred. Studying abroad is popular because most courses are approved by U.S. universities, which makes it easy to transfer credits back to your home university. The sponsoring organization takes care of most of the necessary paperwork for you, including arranging your passport, transportation, housing, course enrollment, and credit transfers. The traditional types of structured study-abroad programs offered by most universities are designed with the American curricula, the American student's needs, and American teaching methods in mind.

Some exchange organizations, such as CIEE, offer highly motivated and independent-minded students the opportunity to study at foreign universities through direct enrollment in that university's courses. Direct enrollment offers total immersion in a country's culture, language, and academic way of life. Participating in classes where you may be the only foreigner from your country can be both challenging and enlightening, offering a very different experience from studying abroad in a traditional program with students of your same nationality. Direct-enrollment programs may also allow you to meet students from parts of the world that you may not otherwise meet in a traditionally structured study-abroad exchange program. Whether you study abroad through a traditional or direct-enrollment program, there are a few things you'll have to prepare for:

- As you read about the different programs, be aware of prerequisites and application deadlines. Deadlines are often six to twelve months before the start of a program, and sometimes language proficiency is required. If you are a degree-seeking student, also make sure that academic credit will be transferable and that the study-abroad program fits into your academic schedule. If you would like to enroll directly in a foreign university, verify that you will receive an official transcript and that your course credits will apply toward your degree. Program details can usually be requested free of charge by phone, fax, or email.

- Apply to several programs in case you don't get accepted into your first choice. Also, inquire about applying any current financial aid you might be receiving to your study-abroad program costs. There are ways to finance your overseas study, such as scholarships, fellowships, and low-interest loans, which I'll discuss in chapter nine.

- You will need to get a passport, if you don't already have one. Sometimes your study-abroad office or exchange program will apply for the passport for you. Otherwise you will have to get your passport on your own. See http://travel.state.gov/passport_services.html for a passport agency near you, and allow two months for processing.

- Apply for an International Student Identity Card (ISIC), which enables you to receive discounts on travel, airfare, and lodging worldwide. The ISIC automatically insures you against sickness and accidents outside of the United States. The Council on International Educational Exchange (www.ciee.org) is the official U.S. sponsor of the ISIC, so it can be obtained through their office. Also, check with your program to see if you will need vaccinations or a visa.

- Depending on where you are going and how long you will be away, you might consider buying a phone card or getting an international cell phone. GoAbroad (see page 128) offers a free membership to eKit, an all-in-one phone/voicemail/free email package (see www.goabroad.ekit.com).

Programs for High School Students

The sooner you can start adding international sojourns to your life, the better. High school is a great time to begin traveling and shaping your future. Take advantage of

A Global Citizen's Perspective

Worlds Apart

by Claire Campbell (clairehmcampbell@hotmail.com)

My first trip out of the country was to Cuba during the interim term of my sophomore year at St. Olaf College in Northfield, Minnesota. I was nineteen. The majority of St. Olaf students travel abroad during interim, as some curricula will not allow semesters abroad. Over 66 percent of my fellow students study abroad at one time during their college career.

My interim class to Cuba was through the History and Hispanic Studies Departments, and was called "Revolutionary Cuba." There were twenty students in the class, plus our professor, who planned the trip. We traveled in Cuba for five weeks. From Havana we flew across the country to Santiago de Cuba, took a bus to Trinidad, Camaguey, Varadero, Guama, and then back to Havana. We attended class meetings once a week, and the rest of the time our classes consisted of guest speakers.

Cuba was such a different world. I was nervous that the Cubans wouldn't like me because I am an American, but it was exactly the opposite. They were so happy that Americans made the effort to learn about their culture and their situation that they spent a lot of time talking to us. The cars and buildings were from the 1950s and everything was in Spanish. I loved every minute of it, even the blackouts, traffic jams, and long bus rides. I stayed with a family for just a day because it is illegal to spend the night in the home of Cubans if you are a foreigner. My host family was so warm and welcoming, and even though we only spent a day together, I cried when I left them. They presented me with gifts upon my departure, which was incredibly moving because Cubans don't have many material items to give in the first place.

We had a party for all of the students and host families on the last day. Some students wanted to go dancing, so I asked my hosts if they would come. They said they'd like to, but the club where people were going didn't allow Cubans, only tourists. It was a really big shock that such laws existed in a country where they stressed equality. I was so irate that something like that existed, and even more irate that some students on my trip went to the club despite this injustice. I stayed with my hosts and we danced at the party.

My experience in Cuba helped me get an internship at a Latin American nongovernmental organization (NGO) in Washington, D.C., the summer after my sophomore year. That experience helped me get an internship at another Latin American NGO the following summer. I was able to get both of my internships because I had traveled to a Latin American country and had become proficient in Spanish.

My second trip abroad was during my junior year of college. I lived in Ireland for four and a half months. This trip was also planned through my school, however the study-abroad program only arranged my admission to the Irish university I attended. They did not provide me with room, board, or transportation to the country. During orientation, I was given a list of potential housing, where to get groceries, and when I needed to be at school. The rest was up to me to figure out. I attended the National University of Ireland at Galway, which has ten thousand students, much larger than I was used to. I was raised in a proud Irish family, and I had always wanted to visit Ireland. The classes I took at the National University at Galway complemented my political science major. I traveled all over Ireland, including Northern Ireland, and to Great Britain, France, Italy, Austria, and the Czech Republic.

After such an amazing experience in Cuba, I expected the same in Ireland, but learned that nothing can top that first international trip. I expected Ireland to welcome me home with open arms. I didn't realize Ireland would have different customs, cultures, and terms. That was a shock—I had expected to settle in comfortably, but I really had trouble adjusting. Although I'm sad to say it, Ireland felt like a warped version of the United States for a little while. I finally became comfortable, but it wasn't like "the motherland" was calling me home. I felt more of an attachment to Cuba, because I connected with the people so well. Academically, Ireland was much easier than I had imagined. I entered the semester prepared to study hard and write papers, but most of the classes only required one paper for the entire term, and we didn't have any exams until the end. The students were really friendly and toward the end of the term I'd see my classmates in the pubs, and we would carry on lively conversations.

My most memorable experience in Ireland occurred during my visit to Belfast. Although Belfast was in the midst of a cease-fire, it was obvious that the Troubles were far from over. People wanted to know my last name to determine whether I was Catholic or Protestant. Race didn't matter there, religion did. As I was touring West Belfast, the Catholic area of the city, I was saddened because it really did look like a war zone. Shutters were closed, barbed wire was everywhere, and murals on the ends of housing blocks depicted how tired and ragged the population had become. In contrast, East Belfast, the Protestant area, was full of tree-lined avenues and neat little houses. The murals there depicted men in ski masks with machine guns, fighting for the cause. There were a lot of happy moments on my trip, but the ones I learned from and grew from are the moments that made me realize just how lucky I am, and how much work there is to be done to ensure that other people feel that way too. This realization motivates me when I'm working eighteen-hour days, and when I think I deserve more than what I've been dealt. I argue with people now who complain that they don't vote because it doesn't matter anyway. If only they could see what people go through to be able to voice their concerns and opinions in other parts of the world.

I think I grew enormously during and after my international experiences. It is hard being on your own in another country, away from those that know you the best. You get to know yourself really well. Traveling also helped me figure out what was really important in my life, and what wasn't worth worrying about. In addition to maturing through travel, I learned to value my family much more. It is sort of strange what you miss when you are out of the country; a lot of people found themselves missing things they didn't normally miss while studying in the country. Some people missed Taco Bell, even though they never ate there; others missed chocolate cake although they rarely ate it. I missed my family. I didn't really miss them when I was away at college in the United States, but I missed them desperately while I was traveling abroad. If I was homesick, or if I was frustrated about something, I could call them and tell them, and they would understand and give me unconditional support. I also came to value the United States itself much more. Although it sounds a little silly, and a little patriotic, I found myself appreciating the freedoms and opportunities we have as Americans. I thought more about our democratic freedoms while in Cuba and about our many opportunities while in Ireland.

If you want to study abroad, first try a program at your school. If you're still deciding on a school, pick a school that has good study-abroad programs. I had always wanted to travel, and I picked my college (St. Olaf) because of its excellent study-abroad programs. College study-abroad programs take care of some of the major issues, like acceptance into the college overseas and orientation for your trip. If you go with a program, you don't necessarily need to go with a group. My trip to Cuba was with a group, but my trip to Ireland was by myself. If you like having things planned out with a schedule to follow, a group trip is the way to go. It is sometimes very easy to travel that way. If you like wandering and discovering, then go without a group.

If you ever have trouble deciding whether to stay or go, GO. You will never regret it, and you will never forget it. I think it is sad if people have the opportunity to study abroad and pass it up because they think they'll miss home. Your real friends will be there for you no matter where you are in the world, and your family will be there for you forever. If friendships dwindle and die, then they weren't going to last anyway. You'll make new friends on your travels, which can be even more exciting because they have so much to offer. Things will still be relatively the same when you get back, and there is always air mail. Whatever you do, go somewhere overseas; I guarantee that you'll be happy you did.

study, work, and volunteer abroad programs, and scholarships, low-interest loans, and student discounts on airfares and lodging. International experience can help you get into a better college or graduate program, learn life skills, develop a sense of independence, prepare you to communicate and live in an increasingly global society, and expose you to life options that you may not find in your home country.

The programs listed below comprise the most reputable organizations that sponsor international exchange students around the world for high school students ages fifteen to eighteen. Summer, academic-semester, or year exchanges where the student lives with a host family and attends a local school are available, as are academic credit and financial aid. All programs comply with the Council on Standards for International Education Travel's (CSIET) standards (see below) for international educational travel programs and are listed in the 2000–01 CSIET Advisory List.

★ AFS (American Field Service) Intercultural Programs

www.afs.org/usa

Intercultural Programs USA
198 Madison Avenue, 8th Floor
New York, NY 10016
Toll-free: 800-AFS-INFO
Phone: 503-241-1578
Fax: 503-241-1653

AFS sends students to over forty-five countries. Students must have a minimum GPA of 2.80. AFS also offers interim programs (offered between semesters) for students over eighteen, and teacher exchanges. Their Web site has a listing of specific colleges that have designated scholarships for students with international experience.

★ AYUSA International

www.ayusa.org

Study-Abroad Department
One Post Street, 7th Floor

San Francisco, CA 94104
Toll-free: 800-727-4540
Fax: 415-986-4620

Students must have at least a 2.75 GPA for academic-year and semester programs and a 2.50 GPA for summer programs. AYUSA International offers some full and partial financial assistance as well as scholarships for hosting international students. Scholarships are available through the Congress-Bundestag program, Kiwanis International, and from corporate and individual donations. AYUSA also provides information on how to raise funds to become an exchange student.

★ CSIET (Council on Standards for International Education Travel)

www.csiet.org

212 S. Henry Street
Alexandria, VA 22314
Phone: 703-739-9050
Fax: 703-793-9035

CSIET is a nonprofit organization committed to quality international educational exchange. It establishes standards for international educational travel and exchange programs for high school students, and monitors compliance with those standards.

★ Organization for Cultural Exchange Among Nations (O.C.E.A.N.) Outbound Program

www.oceanintl.org

2101 East Broadway Road, Suite 2
Tempe, AZ 85282
Toll-free: 800-28-OCEAN
Fax: 480-784-4891

Students are selected on the basis of language proficiency and academic excellence. O.C.E.A.N. has implemented a grant program that helps deserving individuals attain their dreams with financial aid based on their specific needs.

Rotary Youth Exchange

www.rotary.org

One Rotary Center
1560 Sherman Avenue
Evanston, IL 60201
Phone: 847-866-3000
Fax: 847-328-8554

Students must be sponsored by a local Rotary club or district. Students' eligibility is based on scholastic record, linguistic ability, community involvement, potential as a cultural ambassador, and age. There are Rotary clubs in sixty-five countries. For long-term exchanges of one year, the Rotary club usually pays for tuition and provides a modest monthly allowance.

Youth for Understanding (YFU)

www.youthforunderstanding.org

Toll-free: 800-TEENAGE
See the Web site for a YFU office in your country.

YFU is one of the oldest, largest, and most respected student exchange organizations, with over ninety programs in thirty countries. Semester and year programs require a 3.0 GPA, and summer programs require a 2.0 GPA. For most countries, you don't need to know the language. Alumni, volunteers, and host families may be eligible to receive discounts.

Programs for Undergraduates and Graduates

Regardless of your college major, international experience will prepare you for the future in this increasingly global world. There are literally thousands of study-abroad programs for undergraduates. Nearly every college and university in the United States offers study-abroad options and your college or university's study-abroad office should be your first stop for research. You can also research programs via Internet Web sites whose databases are designed to help you narrow down the overwhelming number of options. If you want to study economics in Mexico, Japanese in Japan, or international business in Italy, the best way to locate programs for undergraduates is to do a quick search on CIEE (see page 127), IIE (see page 129), or Studyabroad.com (see page 129). Following are some of the more unique educational exchange programs and institutions whose mission is to provide students with international education and cultural immersion.

★ College Consortium for International Studies (CCIS)

www.ccisabroad.org

2000 P Street, NW
Suite 503
Washington, DC 20036
Toll-free: 800-453-6956
Phone: 202-223-0330
Fax: 202-223-0999

CCIS is a consortium of American and foreign universities that sponsor forty study-abroad programs in twenty-eight countries for undergraduates, and professional development programs for faculty and administrators designed to increase international awareness. The minimum age is eighteen and the minimum GPA to take part in a study-abroad program is 2.5–3.0 on a 4.0 scale. CCIS has a range of courses that can be taken in English or a foreign language. Students will be placed with a host family, but may also stay in a dormitory or apartment. You can search the CCIS database for semester, summer, and short-term programs. Although costs vary per country, if your school is a CCIS member, you can be nominated for financial-aid scholarships.

★ DIS, Denmark's International Study Program

www.disp.dk

DIS North American Office
214 Heller Hall
University of Minnesota
271 19th Avenue South
Minneapolis, MN 55455
Toll-free: 800-247-3477
Phone: 612-626-7679
Fax: 612-626-8009

DIS Copenhagen Office
Vestergade 7
DK-1456 Copenhagen K
Denmark
Phone: (45) 33 11 01 44
Fax: (45) 33 93 26 24

Affiliated with the University of Copenhagen, the DIS Program offers summer, semester, and yearlong study-abroad courses in humanities, international business, architecture and design, medical practice, child development, and Arctic biology (in Iceland). Courses include several European study tours and a six-week academic internship in London is also available. DIS fees cover expenses like local commuting and health insurance, which is better than many study-abroad programs. Costs vary according to program length and several scholarships are available through DIS.

A GLOBAL CITIZEN'S PERSPECTIVE

Studying Abroad in Denmark

by Elizabeth Kruempelmann (ekruempe@hotmail.com)

I first went abroad when I was twenty-one years old. I spent my senior year in college studying international relations and business through an excellent exchange program between my university, the State University of New York at Albany, and DIS, Denmark's International Study Program (www.disp.dk) affiliated with the University of Copenhagen.

My courses in international relations and business were much more stimulating than courses at my home university. My Danish program integrated real-life case studies of companies into our coursework and included cultural study tours to various European countries. For example, my business courses included several weeks of business trips to firms in Denmark, Germany, the former Soviet Union, Poland, and Latvia. We didn't just tour the offices and read annual reports. Our whole class was given the unique opportunity to meet with top managers and discuss their current business challenges. Then, as part of a real case study, our class broke into groups, brainstormed solutions, and presented our ideas to the managers and our professors. The hands-on learning and cultural exposure were definitely experiences I would not have had at my home university.

I also experienced history in the making when the Berlin Wall came down and the ideals of glasnost and perestroika opened up the Eastern bloc to Western influence. These events prompted incredibly enlightening discussions with my professors, who were renowned experts in the fields of Russian and East European affairs. Being in Germany, Russia, and Eastern Europe to experience firsthand the excitement of these world-changing events made such a deep and lasting impression that I promised myself I would live in Europe again someday.

Read more about Elizabeth Kruempelmann's international experiences in chapters two, seven, and eight.

The Institute for the International Education of Students (IES)

www.iesabroad.org

33 N. LaSalle Street, 15th Floor
Chicago, IL 60602
Toll-Free: 800-995-2300
Phone: 312-944-1750
Fax: 312-944-1488

IES works with five hundred colleges and universities to blend the requirements of U.S. universities with the advantages of foreign higher education. IES offers twenty-two unique study-abroad programs in Europe, Asia, Australia, and South America, encompassing a variety of academic fields. Students can take part in field studies, sports, and volunteering, as well as enroll in local university courses or create their own internships. Credits are easily transferable back to your home university. IES believes that financial constraints should not prevent a student from studying abroad, and commits over one million dollars to scholarships and other forms of financial aid.

The International Partnership for Service-Learning (IPS-L)

www.ipsl.org

815 Second Avenue, Suite 315
New York, NY 10017
Phone: 212-986-0989
Fax: 212-986-5039

IPS-L is an incorporated not-for-profit organization that combines academic study with community learning. It serves undergraduates, graduates, and in-service professionals by designing and administering off-campus programs in the Czech Republic, Ecuador, England, France, India, Israel, Jamaica, Mexico, the Philippines, Scotland, and South Dakota (with Native American Indians). IPS-L has established relationships with partner agencies in communities worldwide, thus making it possible for students to serve fifteen hours per week at schools, orphanages, health-care agencies, educational institutions, recreation centers, and community-development organizations. The merging of academic curriculum with service work creates a holistic approach to learning. Students receive credit for their academic work, but not for the community service portion of the program. Applications are due at least two months before the start of the program. Costs vary considerably by country. Summer programs may cost as little as $3,000 USD and year programs may be as much as $17,000 USD. Various types of financial aid can be applied to IPS-L's study programs, including work-study funds. Work-study allows students to work or do community service in exchange for money to pay for classes. Apply for work-study funds through your university's financial-aid office.

★ International Student Exchange Program (ISEP)

www.isep.org

1601 Connecticut Avenue, NW, Suite 501
Washington, DC 20009-1305
Phone: 202-667-8027
Fax: 202-667-7801

ISEP is a nonprofit student-exchange membership organization of 220 colleges and universities in the United States and 35 other countries. Exchanges are offered in 110 locations in Europe, Latin America, Australia, Canada, Asia, and Africa for an academic semester, year, or summer term. Students must apply to ISEP programs through their home universities. Enrollment is open to sophomores through graduate students in a range of fields, including technical areas of study. Students who are proficient in a foreign language can study in that language with local students. Credits can be applied to your degree. The costs of the programs are equivalent to the tuition, room, and board at your home institution, with financial aid being applicable.

 **Schiller International University**

www.schiller.edu

Dept. TA
453 Edgewater Drive
Dunedin, FL 34698
Phone: 813-736-5082
Fax: 813-734-0359

Schiller University offers international study in England, Germany, France, Switzerland, and Spain. Undergraduate and graduate programs include business, hotel management, international relations and diplomacy, European studies, liberal arts, psychology, languages, computer systems management, commercial art, and more. You can earn credit toward your degree at any of their campuses. All study-abroad programs are in English. You can take part on a semester, summer, and year degree program. Costs vary, but financial aid may be available.

★ **The School for Field Studies (SFS)**

www.fieldstudies.org

16 Broadway
Beverly, MA 01915-4435
Toll-free: 800-989-4435
Phone: 978-927-7777
Fax: 978-927-5127

SFS is America's oldest and largest educational institution exclusively dedicated to training and engaging undergraduates in international environmental problem solving. College and high school students can live and study in some of the world's most beautiful and threatened ecosystems while earning college credit for participating in field studies of challenging environmental dilemmas. Summer and semester field-study programs are available in

Australia, British West Indies, Canada, Costa Rica, Kenya, and Mexico. You might investigate new approaches to fisheries management and marine park development on Turks and Caicos Islands, work with Costa Rican locals to develop sustainable resource options, or identify ways to manage endangered sea turtles and marine mammals in Mexico. Courses range from twenty-five to thirty-four students, and costs vary per program. Approximately one-third of the participants receive financial aid.

★ School for International Training (SIT)/World Learning

www.sit.edu or www.worldlearning.org

Kipling Road
P.O. Box 676
Brattleboro, VT 05302-0676
Phone: 802-257-7751
Fax: 802-258-3248

World Learning (founded in 1932 under the name The U.S. Experiment in International Living) is the only international organization dedicated to furthering intercultural understanding, social justice, and world peace through academic and special project entities. The School for International Training (SIT) is World Learning's accredited college. SIT's study-abroad division offers fifty-seven study programs in over forty countries and is one of the world's most comprehensive study-abroad programs. SIT awards sixteen undergraduate credits for the completion of a study-abroad program, which are easily transferable to your home institution. Program fees vary, but range between $10,600 and $14,000 USD. A limited number of scholarships, grants, and financial aid is available, but your best options for financial aid are through your university. SIT offers graduate and degree programs, study abroad, and international exchange, international development, extension and certificate programs, language and intercultural training, teacher education and professional development, and peace building and social justice initiatives.

World Learning's educational and training programs enable participants to develop the leadership capabilities and cross-cultural competence required to advance international understanding, work effectively in multicultural environments, and achieve sustainable development at the community level and on a national or global scale. World Learning programs also encompass high school exchanges through the Experiment in International Living.

★ Semester at Sea

www.semesteratsea.com

811 William Pitt Union
University of Pittsburgh
Pittsburgh, PA 15260
Toll-free: 800-854-0195
Phone: 412-648-7490

Semester at Sea is not just another semester of school; it is a life-altering learning adventure where students study aboard a passenger ship. Semester at Sea is an international program, sponsored academically by the University of Pittsburgh, that takes undergraduate students to the four corners of the world and provides the opportunity for firsthand observation and participation, thus bringing an intensely new dimension to your academic career. The richly varied international curriculum, combined with a profound opportunity for in-country fieldwork, provides an investment in your future. Students are required to participate in an international field program during visits to nine or ten countries in diverse parts of the world. The program is open to all majors and offers approximately seventy courses.

Semester at Sea offers studies in such places as Brazil, China, Egypt, Greece, India, Israel, Japan, Kenya, Morocco, Turkey, Venezuela, and Vietnam. Visit the Great Wall of China, the Taj Mahal in India, or the Great Pyramids in Egypt. Learn about wildlife, ecological issues, and the life of the Masai while on safari in Kenya. Or study the flora and fauna of tropical rain forests and learn about rain-forest destruction while canoeing down the Amazon River. Spend the night in an "untouchable" village in India, visit a black township in South Africa, or interact with local business and professional leaders.

The shipboard campus includes classrooms, study lounges, a library, and a theater, plus two dining rooms, student union, campus store, snack bar, swimming pool, fitness center, and health clinic. Classes meet daily while at sea, with class sizes averaging twenty to thirty students. Port stays of four to six days offer a firsthand look at the societies you will be studying in the classroom. Educational activities on land are designed to complement classroom instruction and other field activities provide opportunities for travel to places of significant historical, cultural, and political interest. Admissions are on a rolling basis and it is recommended that you start the application process twelve to eighteen months ahead of time. You can apply financial aid from your home institution or apply for financial assistance from Semester at Sea. If you have previous experience working on campus as a resident assistant, resident director, or similar position, you might be able to land yourself a paid position on board, as a friend of mine was able to do.

★ Studio Art Centers International (SACI)

www.saci-florence.org

SACI Coordinator
U.S. Student Programs, Institute of International Education
809 United Nations Plaza
New York, NY 10017
Phone: 212-984-5548
Fax: 212-9845325

SACI, centered in Florence, Italy, is one of the most recognized overseas institutions in the fields of studio art, art history, art conservation, and Italian language. Field trips to museums, cathedrals, and monuments in

A GLOBAL CITIZEN'S PERSPECTIVE

Seeing the World through Semester at Sea

by Christina Picardi, ESL teacher (Travelme32@aol.com)

I was raised by parents who have yet to fly on an airplane, so all of my travel experiences are self-motivated. I've always had a curiosity about other people and other cultures, the way they live, their families, their cultural traditions, and their food. During the final semester of my master's program, a colleague introduced me to Semester at Sea. In addition to the students, approximately forty "senior adult" participants join each voyage.

I had just spent six continuous years studying for my undergraduate and graduate diplomas. Taking a few months to travel the world seemed like a great idea, so I applied. I figured I'd get the travel bug out of my system, then could move on with my career and be satisfied. Well, little did I realize that the travel bug does not leave your system, once it gets in! I was accepted for the position of resident director with the Institute for Shipboard Education. The opportunity was a dream come true. My major area of responsibility on the voyage was to work with student ambassadors. This dual role created a network of students who acted as hosts for our international lecturers and visitors. Also, together we planned the voyage's finale: The Ambassador's Ball, a formal event for the entire shipboard community.

We spent an average of fifty days at sea and fifty days in the countries. The voyage began in Nassau, Bahamas, and included the following ports of call: La Guaira, Venezuela; Salvador, Brazil; Cape Town, South Africa; Mombasa, Kenya; Madras, India; Penang, Malaysia; Singapore; Osaka, Japan; Shanghai, China; and Hong Kong. We spent about three to five days in each port. The program also allows extensive travel within each country. For instance, when I was in India, I flew to Delhi/Agra to see the Taj Mahal, and in China I flew to Beijing to see the Great Wall. When the voyage ended in Hong Kong, I continued traveling with a group of students and staff on a postvoyage excursion to Ho Chi Minh City, Vietnam, and Bangkok, Thailand.

During the trip I met and talked with Mother Teresa of Calcutta on the airplane from Ho Chi Minh City to Bangkok; Bishop Desmond Tutu invited the shipboard community to a special mass while we were in South Africa; and some of us attended a rally in honor of Nelson Mandela.

The Semester-at-Sea experience not only made me more culturally aware, it made me appreciate what we have in the United States. As an ESL teacher, I try to get to know my students as students as well as people—where they come from, their traditions, and heritage. Semester at Sea has made me a different sort of traveler. Now I stay at a Pousada rather than a Hyatt Regency, and I might rent a car rather than take a train. I initially thought Semester at Sea would cure my travel bug, but instead it has made it grow stronger. I have traveled to more than forty countries, and I hope this list continues to grow.

Read more about Christina Picardi's international experiences in chapter four.

Rome, Milan, Bologna, and other important art centers in Italy are an integral part of the academic programs. Artists, curators, museum directors, and other professionals present a series of lectures to students during the course of the study program. You can take individual courses, study for a semester or a year, or get a two-year diploma, three-year postbaccalaureate certificate, or a master's degree in fine arts.

INTERNATIONAL BUSINESS (MBA/MIM) PROGRAMS AND EXECUTIVE EDUCATION

What do you do if you've been out of college for a while, have relevant work experience, and decide that it's time to start seeing the world? You make working abroad your new goal. In fact, maybe you've already jumped ahead to the working abroad chapters and ended up back here. If you are a career changer or if you just need to internationalize your current career direction, an international business degree is a great place to start.

An international master's degree in business will help you develop the business knowledge, language skills, and international work experience you lack. Many people interested in eventually working abroad consider an international MBA or MIM (Master of International Management) to be the first step to an international career, especially if you want to work at a Fortune 500 company. One of the most important aspects of gaining an international management or business degree is the contact you will have with international students, professors, and business professionals. These are the people who can help you build your international career, so be sure to stay in touch.

Prerequisites vary per program, but many require a few years of previous business experience. Some of the more demanding programs expect previous experience living overseas and fluency in at least one to three foreign languages. The Graduate Management Admission Test (GMAT) will also be required.

International graduate business programs don't come cheap. Scholarships and loans are generally available if you will be paying for this coveted degree yourself. Most reputable schools have good career placement offices that will help you obtain appealing job offers (with salaries to match), that will help you pay off your B-school debt. However, if you work for an international company, see if it will foot the bill for your international education. If so, you will probably have to agree to continue working at that company for a certain amount of time after you get your degree.

In addition to offering international graduate degrees in business, many universities and corporations organize executive and certificate programs for professionals. For example, The American Graduate School of International Management (see Thunderbird below) offers a Global Executive Development Asia Series (Thunderbird MIM for Executives in Asia) in cooperation with the Acer Group, covering topics such as globalization, marketing, leadership, and more. The first and last modules are presented at the Thunderbird campus in Arizona, while the remaining modules are held over fourteen months at Acer's Aspire Academy in Long-tan, Taiwan. For more about this program, see www.t-bird.edu/execed/leadership.

Some of the best-known and most respected international business programs at the graduate level are listed below. Their graduates can be found working in top-level business positions at international companies around the globe.

★ ESCP-EAP European School of Management

www.eap.ccip.fr

79 avenue de la République
75543 Paris Cedex 11
France
Phone: (33) 1 49 23 27 90
Fax: (33) 1 49 23 22 25

EAP offers a European MBA, a European Master in Management (EMIM) and a range of executive programs. Campuses are in Paris, Oxford, Berlin, and Madrid. The MBA program is either twelve months full-time in Paris or eighteen months part-time in Madrid (both programs are presented in English). The EMIM is a generalist management course where students spend a year in three out of four countries: France, the U.K., Germany, or Spain. Each course is taught in the local language. The EMIM also includes three-month company placements in each country. Tuition for the full-time

MBA is 20,000 Euros (about $21,800 USD, subject to change), which includes compulsory seminars in Brussels, Warsaw, Berlin, and Madrid.

★ **INSEAD**

www.insead.edu

European Campus
Boulevard de Constance
F-77305 Fontainbleau Cedex
France
Phone: (33) (0) 1 60 72 40 00
Fax: (33) (0) 1 60 74 55 00

Singapore Campus
1 Ayer Rajah Avenue
Buona Vista
138676 Singapore
Phone: (65) 799 5388
Fax: (65) 799 5389

INSEAD is one of the world's largest and most recognized graduate business schools, offering a one-year intensive master's degree in international business administration.

Campus exchanges are encouraged between INSEAD's two campuses in Fontainebleau and Singapore. In the 2000–01 year, there were 726 MBA students, 6,000 executive program participants, and 21,000 alumni worldwide. For the year 2000, individuals who paid their own way had to dish out 35,500 Euros ($38,150 USD) and companies paying for their employees would be charged the full tuition price of 47,500 Euros ($51,230 USD). It is recommended that individuals budget about 17,225 Euros ($18,775 USD) for accommodation, books, and other expenses. INSEAD scholarships and loans are available.

★ International Institute for Management Development (IMD)

www.imd.ch

Chemin de Bellerive 23
P.O. Box 915
CH-1001 Lausanne
Switzerland
Phone: (41) 21-618-01-11
Fax: (41) 21-618-07-07

IMD is rated among the top three business schools in Europe, and offers a full-time and executive MBA. The full-time MBA program runs from January to December and includes about eighty students from over thirty different countries. The MBA program emphasizes general management in a global world, where the students are expected to interact, drawing on their previous work experiences. The typical MBA student is a high-potential professional between twenty-six and thirty-five years old, has at least three years of work experience, and speaks three languages. For the year 2002, the MBA program costs 49,500 Swiss Francs ($80,190 USD). If your employer is picking up the bill, it will be 72,000 Swiss Francs ($116,640 USD). IMD offers scholarships and loans.

★ Thunderbird, The American Graduate School of International Management

www.t-bird.edu/default.asp

15249 N. 59th Avenue
Glendale, AZ 85306
Phone: 602-978-7100
Fax: 602-439-5432

Thunderbird, located in Arizona, offers a Master of International Management (MIM), that includes a curriculum in world business, international studies, and modern languages. In addition to the MIM degree, programs are offered in

MIM-Latin America, MIM for Executives in Asia, and dual MIM/MBA degrees. If you spend some time investigating different programs, you may hear about the infamous Thunderbird network of over 29,000 alumni working in 135 countries. From my experience living abroad, many of the professionals I met with master's degrees in management came from Thunderbird. I've seen the Thunderbird network in action and can attest to its worldwide reputation for producing truly global-minded managers. Thunderbird offers full-time degrees and certificates, degrees for working professionals, executive education, and overseas programs. Costs vary per program and financial aid is available.

INTERNATIONAL SCHOLARSHIPS, FELLOWSHIPS, AND RESEARCH GRANTS

There are several ways for you to learn more about your professional specialty overseas. In addition to study-abroad options, you may be able to arrange graduate or postgraduate work in your field through scholarships, fellowships, and research grants. Contrary to popular belief, scholarships and fellowships are not just for academics. They can also be available to working professionals who are not enrolled in an academic program. There are thousands of grants offered through universities, government and nongovernment organizations, professional associations, and the corporate world. If you are a graduate student, check with your university's study-abroad office. Some fellowships are very specific to a course of study, and others have stringent prerequisites. But, the good news is that there are lots of easier-to-come-by options with fewer restrictions on the type of research you undertake. For example, the Institute of International Education (IIE) offers Fulbright scholarship and fellowship programs to students, scholars, and business professionals to study, lecture, give seminars, and conduct research on a variety of topics (see below). See chapter nine for scholarships, fellowships, and grants that can be used as financial aid.

★ **THE FULBRIGHT SCHOLAR PROGRAM (FOR POSTBACCALAUREATE DEGREES)**
www.iie.org/cies

COUNCIL FOR INTERNATIONAL EXCHANGE OF SCHOLARS (FOR POSTDOCTORATE SCHOLARS)

www.cies.org

3007 Tilden Street NW, Suite 5L
Washington, DC 20008
Phone: 202-686-4000
Fax: 202-362-3442

The Fulbright Scholar Program, a program of IIE administered by CIES, is the main scholarship program sponsored by the U.S. government. Contrary to some of the myths you might have heard about Fulbright, scholarships are available not only to scholars, but to students, lawyers, journalists, musicians, filmmakers, scientists, engineers, and artists. You could qualify for the various Fulbright programs to study, teach, lecture, and do research in one of 140 countries. Some programs require a foreign language, but others do not. As the program was designed to promote self-development through international experience, the program allows the participants to create their own research agendas or special projects. For example, you might study contemporary artistic expression in India, research women's rights in Chile, or be directly involved in cancer research in the U.K. Full grants are offered and include all transportation, living, and research expenses, and health and accident insurance. Travel grants to Germany, Hungary, and Italy are offered to supplement a non-IIE–sponsored scholarship award. Foreign and private grants are also available. Alternatively, you could apply for a grant to teach English, or a business grant to research and intern abroad. On the Web site you can search grant summaries by country. Eligibility requirements vary per program, but may not be as strict as other scholarship programs. Programs range from two months up to one year.

★ **National Security Education Program (NSEP)**

www.iie.org/nsep

809 United Nations Plaza
New York, NY 10017

Toll-free: 800-618-NSEP
Phone: 202-326-7697

NSEP, part of the Institute of International Education, exists to help Americans understand foreign cultures, strengthen U.S. competitiveness in world markets, and enhance international cooperation and security. Scholarships are available to students who study less common world languages, cultures, and issues that are critical to U.S. national security, such as sustainable development, environmental degradation, global disease, and economic competitiveness. Seventy-nine world regions and forty-two languages are considered for scholarships, excluding Western Europe, Canada, New Zealand, and Australia. A critical factor in the decision process is how closely your proposed study is related to your academic and career goals. Scholarship awards are for a minimum of one term, but preferably for one full academic year. NSEP requires that scholarship recipients seek employment with a federal agency involved with national security.

★ Office of English Language Programs

http://exchanges.state.gov/education/engteaching/index.htm

U.S. Department of State (Annex 44)
301 4th Street SW, Room 304
Washington, DC 20547
Phone: 202-619-5869
Fax: 202-401-1250

This office is responsible for U.S. government English-teaching support activities outside of the United States and offers three grant possibilities: (1) The English Teaching Fellow (ETF) grant recruits qualified English as a Foreign Language (EFL) teachers to teach at universities around the world. (2) The EFL program places experienced EFL teacher trainers and English as a Second Language (ESL) instructors in positions in Eastern and Central

The Ins and Outs of Fulbright Scholarships

by Rebecca Falkoff, writer and teacher living in Paris (Rebecca@falkoffs.com)

After six years of stumbling over *salut* in the classroom I decided to give French a rest and fulfill my college language requirement by spending a summer studying in Florence, Italy. It was the first time I experienced living abroad and I loved it. I loved the constant intellectual stimulation of the language and the cultural differences.

Living in Italy, I was struck by the prominence of immigration issues in political and social discourse. Like many Americans abroad, I was accustomed to an ethnically diverse and integrated society, and was startled at how different it seemed in Italy.

When I returned to the United States I began to research the subject and became particularly interested in the way Albanian immigrants are represented in Italian film and literature. For my research proposal on the subject, I was awarded the Amici Prize, an $800 USD grant given by the University of Pennsylvania's Italian Studies Department to students studying in Italy. I studied in Padua, Italy, that spring, and continued this research.

As I began my final year at the University of Pennsylvania, I knew that I wanted to return to Italy upon graduation. I looked into funding for graduate study in Italy and realized that a Fulbright grant would be ideal. I applied for a Fulbright grant to study Albanian immigration to Italy in Padua, and was selected as a finalist. The next year I applied again, this time to study literature written by recent immigrants to Italy in Rome. That year I was selected as an alternate. Although I was not awarded a Fulbright grant, I went to Rome on my own and audited a course on literature in Italian by foreigners.

Basically, a Fulbright grant allows you to spend a year pursuing a specific interest in some aspect of another culture. Almost one thousand Fulbright grants are offered each year in more than one hundred countries. They are available to people with a variety of different backgrounds, from recent college graduates to professionals and artists. The grants generally cover all expenses, including round-trip transportation, tuition, book and research allowances, maintenance for the academic year, and health insurance.

The Fulbright application is relatively straightforward, consisting of only two main essays—a Personal Statement and a Statement of Proposed Study or Research. The application also requires three recommendations written by professors or colleagues. Depending on the country and the project, a language recommendation, which attests to your language ability, may also be required. Slides or other secondary materials may be necessary for art projects.

The Statement of Proposed Study should be detailed and precise. It needs to convey to the Fulbright committee that you are ready to set off for a year of independent study. According to Clare Cowen, the advisor for graduate scholarships abroad at the University of Pennsylvania, the Statement of Proposed Study is the most important part of a Fulbright application. "A good, solid proposal is a very good start," she says.

The project should be rigorous and academically sound, but most importantly, feasible. "I think the worst thing a student can do is to have a proposal that requires using unavailable resources like libraries that haven't been open for years, or archives that are accessible only to a king or queen," Cowen says.

In addition to developing a viable project, you need to seek out the contacts that would be necessary to complete it. This involves establishing some affiliation, usually with a scholar or a university, library, or other institution in the country where you hope to study. The affiliation is a loosely defined but integral part of a Fulbright proposal. Ideally, you should obtain a brief letter from a scholar or professional in your field.

There are two types of applicants for U.S. Student Program grants: those who apply through an academic institution in the United States, and those who apply on their own. If you are currently a college senior, you must apply for the grant through your college. You can obtain the application from the Fulbright program advisor on campus. Other applicants should request the application directly from the Institute of International Education (IIE) headquarters (see page 129).

There are no hard and fast rules about who is eligible for a Fulbright grant. Some universities suggest that only students with GPAs above 3.5 apply, but IIE has no such requirement. It is far more important that you demonstrate a commitment to your research than that you ace Econ 101.

The odds of being awarded a Fulbright are not bad. The most difficult aspect is not so much the competition as the proposal itself. In the 2000-01 competition, there were a total of 4,112 applicants for the U.S. Student Program and a total of 960 grants awarded. That means that about 25 percent of the applicants were awarded grants. Some countries are more competitive than others. For a country-by-country breakdown of the applications received and grants awarded, check out the Institute of International Education Web page (see page 129).

For me, the most frustrating thing about applying for a Fulbright was the endless wait that ensued. The applications are due in October and you find out in January whether you are a finalist. If you are selected as a finalist, you will have to wait until May, when the winners and alternates are announced.

The first time I applied for a Fulbright I was selected as a finalist for my proposal on Albanian immigration to Italy. As a finalist, I made the mistake of brazenly expecting to win a grant, so of course I was devastated when I found out in May that I did not.

My advice for anyone applying for a Fulbright grant is to develop a plan B and be prepared to not win. Because it is so hard for recent graduates to find entry-level positions abroad, developing a plan B is no small feat.

Upon graduating I did not immediately return to Italy. Instead, I moved home to Boston and worked for a Web site. I was happy there, but I missed Italy, and I realized that the longer I stayed away, the harder it would be to go back. I applied for a Fulbright grant a second time, but decided to take my own advice and go to Rome, with or without the Fulbright.

I saved up money, took a certification class to teach English as a Foreign Language, and left for Rome. Once I got there, I audited a course at *Università degli studi di Roma La Sapienza* with a professor I had contacted about the Fulbright. And so, by the time I learned that I was an alternate and would probably not get a grant, I was already living *La Dolce Vita.*

Europe, Eurasia, the Caucasus, and Central Asia. To be eligible, you must be a U.S. citizen with an M.A. or Ph.D. in Teaching English as a Foreign Language/Teaching English as a Second Language (TEFL/TESL) and considerable teacher training experience, preferably overseas. Both ETF and EFL program participants receive a fixed stipend, cost of living allowance, round-trip travel, and insurance. (3) The Short-Term English Language Specialist Program recruits TEFL/TESL and applied linguistic professionals for two- to six-week speaker assignments overseas. Participants receive an honorarium, round-trip travel, in-country travel, and other expenses.

★ Robert Bosch Foundation Fellowship Program

www.cdsintl.org/interns

871 United Nations Plaza, 15th Floor
(First Avenue at 49th Street)
New York, NY 10017
Phone: 212-497-3500
Fax: 212-497-3535

American professionals ages twenty-three to thirty-four working in business, economics, journalism, mass communications, law, political science, public affairs, and other fields, can participate in a nine-month executive-level internship in Germany. The program includes seminars in Berlin, Strasbourg, Paris, and Brussels. Language proficiency by the start of the program is required.

★ Rotary Foundation: Ambassadorial Scholarships

www.rotary.org/foundation/educational/amb_scho/index.html

One Rotary Center
1560 Sherman Avenue
Evanston, IL 60201
Phone: 847-866-3000
Fax: 847-328-8554

The Rotary Club awarded more than twelve hundred scholarships in 2000–01. Recipients from sixty-nine nations study in sixty-four different countries through grants totaling $26 million. The Ambassadorial Scholarship Program, with three types of scholarships (described below), is the world's largest privately funded scholarship program. All applicants must have already completed at least two years of college-level course work or have an equivalent amount of professional experience. See the Web site for other eligibility criteria, which vary per scholarship. In addition, teaching grants are available to teach in developing countries.

- Academic-Year Ambassadorial Scholarship
  The Ambassadorial Scholarship funds up to $25,000 USD to study abroad for an academic year. The Ambassadorial Scholarship is the most common type of scholarship offered, and in 2000–01 the Rotary Club awarded one thousand of these scholarships.

- Multi-Year Ambassadorial Scholarship
  The Multi-Year Ambassadorial Scholarship provides $12,000 USD per year for two to three years of degree study overseas. In 2000–01, 150 scholarships were awarded, mostly by Rotary districts in Japan and Korea.

- Cultural Ambassadorial Scholarship
  Up to $12,000 USD for three months or $19,000 USD for six months is awarded for intensive language and cultural training, including homestays. Cultural Ambassadorial Scholarships are available to study Arabic, English, French, German, Hebrew, Italian, Japanese, Korean, Mandarin Chinese, Polish, Portuguese, Russian, Spanish, Swahili, and Swedish.

- Rotary Grants for University Teachers
  These grants are available for higher education teachers to teach at colleges and universities in developing countries. You can apply for a three- to five-month teaching grant of $12,500 USD or a grant for $22,500 USD

to teach for six to ten months. Forty-seven grants were awarded in 1999–2000.

INTERNATIONAL EXCHANGES FOR TEACHERS AND EDUCATIONAL ADMINISTRATORS

Teachers and educational administrators have a large variety of learning-abroad options at their disposal. One of the most popular is a teaching exchange. If you are a full-time teacher with at least three years of teaching experience, you can trade teaching jobs with a teacher from a different country. You might even trade homes and cars, making the move abroad a very easy transition. Teaching exchanges are available at a minimal cost to you, and you continue to get paid as you would in your home country. It can take up to sixteen months to find an appropriate teaching match in your country of choice, so apply as early as you can.

The first step is to choose the exchange program (see resources below), then get your application, references, and résumé in order. When you apply for an exchange, you will have to creatively "sell" your school, community, house, and geographical area to foreign teachers who may potentially want to trade places with you. The more appealing your situation sounds to someone else, the better chance you have at finding an exchange. Once you receive an exchange offer, you have to decide within a week or two whether you want to take it or leave it. Some people have arranged a perfect match, but more likely than not, you will have to compromise something, like a specific location.

Here are some of the most popular programs and resources for teacher exchanges.

★ **Canadian Education Exchange Foundation (CEEF)**

www.ceef.ca

250 Bayview Drive
Barrie, Ontario L4N 4Y8
Canada
Phone: 705-739-7596
Fax: 705-739-7764

CEEF is a nonprofit exchange organization for both students and teachers. CEEF works with the Canadian Ministry of Education and Training, government agencies, and government-sponsored exchange authorities in other countries to arrange teacher exchanges at the elementary and secondary level. To qualify, you must be an exemplary teacher with five years of teaching experience. Principals and administrators can also apply. Canadian teachers will trade teaching positions for a one-year period, retaining their salary, benefits, and seniority. Teachers are expected to exchange their house, or provide appropriate living arrangements for the foreign exchange teacher. There is a nonrefundable application fee of $107 CDN ($67 USD). Upon confirming your teaching exchange and finding an appropriate match, you will have to pay an additional $321 CDN ($201 USD).

★ **Fulbright Teacher and Administrator Exchange Program**

http://grad.usda.gov/International/ftep.html

Graduate School USDA
600 Maryland Avenue, SW
Suite 320
Washington, DC 20024
Phone: 202-314-3520
Fax: 202-479-6806

Elementary through four-year college teachers and administrators can apply to one of the four options offered by the Fulbright Teacher and Administrator Exchange Program. Spend a semester or year in a direct teacher exchange, six weeks in an administrator exchange, eight weeks in a team-teaching initiative in Brazil, or a summer in a classics seminar in Italy. Placements are determined by the subject and level you teach. Applicants need to be currently employed with at least three years of full-time teaching or administrative experience. The program is open to special education teachers, guidance counselors, school psychologists, librarians, nurses, and media specialists.

You don't necessarily have to speak a foreign language to apply as two-thirds of the participating countries do not have a foreign-language requirement. Your regular salary will be paid by your school district during the exchange. Families may accompany applicants.

Global Teaching Exchange

http://www.teachingexchanges.com

The Works Corporation
4903 Denton
Boise, ID 83706

The Global Teaching Exchange is a great Web site for teachers from all countries looking for teaching exchanges and wanting to network with other exchange teachers. Their motto—"teachers helping teachers with teaching exchanges"—is what it is all about.

EXCHANGES AND TRAINING ABROAD FOR PROFESSIONALS

Whether you are a social-service professional, facilities specialist, or jazz musician, there are exciting opportunities for you to participate in cultural exchanges in a variety of fields. The options for learning about your profession overseas may not be common knowledge, but they *do* exist if you know where to look. Professional exchanges take the form of traveling seminars, fellowships, grants, or other officially sponsored exchanges between government and nonprofit groups. You may also discover global cultural exchanges and training through professional organizations in your field, government programs, or independent agencies. Here are a few places to start your research.

The Council on International Fellowship (CIF)

www.cifinternational.com or www.cifusa.org

Council of International Programs USA
1700 E. 13th Street, Suite 4SE
Cleveland, OH 44143

Phone: 216-566-1906
Fax: 216-566-1490

CIF is a private, voluntary, nonprofit organization whose goal is to promote human understanding through cultural, educational, and professional international exchanges. CIF's programs are mainly for human services professionals, like social workers, care workers, special schoolteachers, youth workers, and psychologists. (Council of International Programs—CIP—is CIF's U.S. affiliate, promoting professional exchanges in the States and offering programs for a wider range of professions.) CIF's international exchange programs include an international group of at least four people, an orientation seminar, field placement on an observation basis, and accommodations with a host family. Programs range from one to eight weeks in Austria, Hungary, Finland, France, Germany, Greece, India, Israel, Italy, Slovenia, the Netherlands, Norway, Scotland, Sweden, Turkey, and the United States (in four-, six-, and twelve-month durations). All programs are in English, except one that is in French. The registration fee, which may be free for some countries or up to $1,000 USD for others, is the total cost to participate in the program.

★ Cultural Programs for Arts Professionals and Educators

http://exchanges.state.gov/education/citizens/culture/

Bureau of Educational and Cultural Affairs
U.S. Department of State
SA-44, 301 4th Street SW, Suite 568
Washington, DC 20547
Phone: 202-619-4779
Fax: 202-619-6315

These cultural programs are strictly for professional American artists, filmmakers, musicians, and other specialists of the arts (not for students or for general adult study). Through cultural exposure, participants enrich their work, enhance their international reputations, and form partnerships with overseas institutions. To promote cultural diversity in the arts, participants can choose programs in literature, cultural preservation, and conservation; initiatives that focus on intellectual property rights; and the use of the arts to promote social issues, such as the role of women in society, drug prevention, and environmental protection. See the Web site for details on these programs and others—for jazz singers, American cultural specialists, international visual and performing arts festivals, film service, creative arts exchanges, and museum partnerships.

★ People to People

www.ambassadorprograms.org

501 East Armour Blvd.
Kansas City, MO 64109
Phone: 816-531-4701
Fax: 816-561-7501

People to People arranges scientific, technical, and professional exchanges for groups of ten to forty-five professionals who work in the same field. In traveling seminars that last ten to fourteen days, delegates and their overseas counterparts exchange information, ideas, and experiences. The program includes visits to institutions, businesses, and schools to learn firsthand how issues in specific fields are being addressed internationally. Areas covered include economics, international relations and business, literature, education, health care and social services, documentary film, history, theater, and more. Environmental technologists have met with Chinese officials to discuss protecting the world's environments, literacy specialists have traveled to South Africa to investigate the challenges faced by their colleagues there, and facilities managers have visited Australia for discussions and site visits. A significant amount of time is spent in organized briefings, conferences, inspection visits to facilities, or informal gatherings, allowing people to network with colleagues who are important to their professional growth.

★ **ROTARY INTERNATIONAL**
www.rotary.org/foundation/educational/gse/index.html

One Rotary Center
1560 Sherman Avenue
Evanston, IL 60201
Phone: 847-866-3000
Fax: 847-328-8554

Rotary International's Group Study Exchange program pairs professionals from different countries for the purpose of exchanging vocational information. Rotary districts choose a team consisting of four non-Rotarians and one Rotarian leader, all from different professions. The team spends four to six weeks studying the host country's institutions, economy, and culture, as well as observing how their specific professions are practiced. Applicants must be employed full-time, have at least two years of professional experience, possess some knowledge of the language of the country they are visiting, and be between twenty-five and forty years old. Upon return, participants are expected to share their experiences with other Rotarians or civic organizations. More than five hundred paired exchanges occur every year. Interested parties should contact their local Rotary club for deadline and application information.

In addition to professional exchanges, opportunities to attend international workshops, conferences, and seminars may be available through your employer, professional organization, or local chamber of commerce. There are also many international associations devoted to promoting certain fields of research and education. For example, if you are a consultant, trainer, human resources professional, manager, or scholar interested in cross-cultural and diversity issues, you may want to attend international conferences and events sponsored by SIETAR—the Society for Intercultural Education, Training, and Research (www.sietarinternational.org). If you are a business professional interested in researching the relationship between business, government, and society, you might consider joining the International Association for Business and Society (www.iabs.net). You can expand your professional knowledge and skills, build contacts, and gain recognition by joining international professional organizations and taking part in their activities. See chapter four (Educational Travel) and chapter six (Volunteering Abroad) for other ways professionals can partake in overseas seminars, workshops, and training.

FOREIGN LANGUAGE TRAINING

Whether you want or need to certify your language ability to study at a foreign university, translate, teach, or conduct business in a foreign language, official language certification can be particularly important and oftentimes required. Certification proves to employers that your language skills have been tested objectively and that you have achieved an accepted level of fluency. Many countries offer officially recognized exams and certifications. Check with embassies, consulates, and language schools to find out what kinds of certifications are offered for your language of choice.

Spanish Language Certification

★ Diploma de Español como Lengua Extranjera/Diploma for Spanish as a Foreign Language (D.E.L.E.)

This is the only certificate nonnative speakers can obtain to verify their Spanish-language ability that is officially recognized by the Spanish Ministry of Education and Science. D.E.L.E. has three levels. Exams take place in May and November and the deadline to enroll is one month prior to the exam. To take the D.E.L.E. exam, contact the Spanish embassy, consulate, or Instituto Cervantes (http://users.rcn.com/cervante.interport//wem3.html).

Spanish for Business and Tourism

★ The Spanish for Business and Tourism Exam

This exam is officially recognized by the Chamber of Commerce in Madrid. To find out about the exam, see the Don Quijote Web site listed below for an institute near you, or contact the nearest Spanish embassy or consulate to find out where you can take the exam in your city.

Note: Preparation courses for D.E.L.E. and Spanish for Business and Tourism are given at Don Quijote institutes (www.donquijote.com/english/courses.official.asp) in Barcelona, Madrid, Granada, Seville, and Salamanca. To take the exams in a country other than Spain, contact the nearest Spanish embassy or consulate.

French Language Certification

★ The Paris Chamber of Commerce and Industry

The Paris Chamber of Commerce and Industry has developed a structure for certification in Business and Professional French, which includes seven exams:

1. The Practical Certificate in Business French
2. The Higher Diploma in Business French
3. The Advanced Diploma in Business French
4. The Certificate in Secretarial French
5. The Certificate in French for Tourism and Hotel Occupations
6. The Certificate in French for Scientific and Technical Positions
7. The Certificate in Legal French

The exams are administered twice a year at six hundred testing centers in ninety countries around the world. Check with the French embassy or consulate for a testing center near you and for the dates of the exam.

German Language Certification

★ The Goethe Institut

www.goethe.de/U.K./ney/enindex.htm

Exams for German are internationally recognized and can be accepted as entrance qualifications to German universities. Located in ninety-three countries, the Goethe Institut offers seven different types of German language exams.

- Three general exams are offered to certify beginning (Zertifikat Deutsch–ZD), intermediate (Zentrale Mittelstufenprüfung–ZMP), and advanced (Zentrale Oberstufenprüfung–ZOP) levels.

- Kleines Deutsches Sprachdiplom (KDS) is accepted by private and public employers as evidence of a high level of German proficiency and exempts students from taking the language proficiency exam to enter a German university.

- Grosses Deutsches Sprachdiplom (GDS) requires near-native level fluency. In certain countries the GDS is proof of language ability necessary to train teachers of German. It also serves as proof of language fluency for teachers within the European Union who want to teach in Germany.

- Zertifikat Deutsch für den Beruf (ZDfB) focuses on language used in business, although the level of German required is similar to the ZD (beginning) level.

- Prüfung Wirtschaftsdeutsch International (PWD) is equivalent to the ZMP (intermediate) level, and includes business and technical knowledge.

CHAPTER SIX

Volunteering Abroad: Lending a Hand While Learning

Never doubt that a small group of committed citizens can change the world. Indeed, it's the only thing that has.

MARGARET MEAD

Volunteering abroad is a unique and generous way to experience the world by traveling, learning, and helping others. Volunteering abroad gives you a chance to learn about other people, new cultures, and different ways of life, while making a significant contribution to a local community, world cause, or environmental issue. Many volunteer opportunities involve working on projects around the world for a few weeks to several years. Although some programs are designed specifically for student volunteers, many of the programs are open to people of all ages, from teenagers to retirees.

It is not unusual to experience a major and profound change in the way you think about yourself, your country, and your worldview after participating in a volunteer program overseas, especially if this is your first time abroad. You will develop an ability to work in diverse international groups, increase your level of sensitivity, and become more open to other ways of thinking.

Today's volunteer programs have ambitious goals and missions. They involve topics of cultural appreciation, environmental awareness, economic sustainability, and social change as well as more traditional causes such as world poverty, AIDS, and illiteracy. Expeditions, internships, trainee programs, and teaching are often classified under volunteer programs, and focus on helping communities in less developed countries improve the quality of life through projects like building bridges, conserving ancient monuments, and protecting natural resources. Although volunteering implies giving of yourself to help others, it serves other purposes as well. Many volunteer organizations provide a simple avenue for participants to learn about a culture, practice a language, gain some work experience, and travel around a region, thus promoting the general mission of cross-cultural understanding and world peace.

Volunteering is a frequently overlooked alternative to learning and working abroad. However, the merits of volunteering make it an inexpensive way to obtain college credits or get international work experience. Volunteering can also be essential to your future career prospects especially if you are interested in working in the nonprofit sector.

Although some volunteer programs require that you pay your own way, you don't have to foot the bill alone. Several organizations that sponsor volunteers, such as the Rotary Club, which offers scholarships, will help volunteers raise funds through contributions, raffles, and other fund-raising means. Depending on how and where you want to volunteer, you may be able to find expenses-paid programs, like the Peace Corps or Concern America, which include a small stipend. In exchange for your time and work, many organizations provide job-specific training, work experience, academic credit, and room, board, and in-country expenses.

How is volunteering abroad different from working overseas? Volunteer programs are not tours or vacations, so you must be willing to do what is needed to help the community or cause. Volunteers need to be open-minded, flexible, ready to learn and work, motivated, and enthusiastic. Make no mistake, volunteering is work. One of the biggest differences between participating in a volunteer program and participating in a work-abroad program is the absence of a steady paycheck. Volunteers donate their time, energy, labor, and commitment—many times at their own expense—to further a cause they believe in, or to gain skills and experience for academic or professional ambitions. All-expenses-paid volunteer opportunities exist for professionals in almost every field, whether you are a business manager, psychologist, or social worker.

The Benefits of Volunteering Abroad

Wondering how volunteering can benefit you? Let's take a look at some of the primary ways volunteering can help you and how you can get the most out of your experience.

LEARN FROM THE SOURCE

Learn firsthand about a community, its culture, the people, and the issues they face. In many cases you will be living among the locals, quite possibly with a host family, which means you'll be living like the locals—with or without hot water, heat, and the comforts you might be used to. You may even be eating exotic foods like fried insects and other local treats. There's no better way to get a sense of a place than by spending some time there and living within its normal conventions.

PROMOTE A CAUSE

Whether you raise money for your trip or pay your own expenses, you know your contributions are being used to further the cause—not to make a profit. The goals of most volunteer programs reflect an organization's overall mission, which could be peace, cooperation, international understanding, justice, tolerance, or just generally making the world a better place.

APPLY YOUR SKILLS

As a volunteer you will use your special skills and knowledge in an unconventional setting. If you come from an engineering background, you might help design and construct a local bridge. If you have business skills, you could help local women set up a business to sell their handicrafts. If you studied medicine, you may work and train other health-care workers. Some volunteer expeditions, internships, and research programs may involve learning new skills, such as scuba diving, foreign languages, or methods of restoration. Either way, volunteering can provide you with résumé-building experience that reflects your unique strengths and new global skills.

LIVE AND WORK OFF THE BEATEN PATH

Many volunteer programs involve living in remote and unique places that you might not ordinarily visit or have access to as a tourist. It is common for volunteers to live in the community in which they are working. This means you may stay at a resident's house as a guest, at a camp, or in a hotel or pension. Your meals will generally be eaten with other project members, your host family, or community sponsors.

PARTICIPATE IN A CROSS-CULTURAL EXCHANGE OF IDEAS

During your volunteer experience you will no doubt come into contact and exchange ideas with your group and the local people. Although group sizes vary, group interaction is an essential element of any program, and one that will aid in your personal growth and development. Be prepared to spend lots of time in close contact with a group of people who are there to work toward a common goal. The group could consist of volunteers from other countries as well as local organizers, thus creating an environment for enriching cross-cultural exchanges.

VOLUNTEER WITHOUT PREVIOUS EXPERIENCE

Generally there are no prerequisites for volunteering. The exceptions would be business and health-care volunteer programs, where a graduate degree and previous experience may be required. Many of the programs do not require facility in a foreign language. However, if you speak another language or have other knowledge you would like to use, such as a diving certification or photography experience, ask the organization how your skills can best be integrated. Most programs supply background information and training for each project.

EARN COLLEGE CREDIT

Depending on the duration and the nature of the volunteer program, you may be able to arrange college credit for your participation since many volunteer programs are organized through college and university study-abroad offices. If you would rather arrange your own volunteer assignment, you may still be able to get college credit or independent study credits. Check with your advisor.

SAVE MONEY ON YOUR TAXES

The volunteer organizations listed in this guide are nonprofit organizations. Usually their main sources of funding are through program fees and contributions. Costs are divided up to pay for administrative overhead, program development, field expenses, participant coordination, and recruitment. One big advantage of volunteer programs is that the program costs and airfare may be tax deductible for U.S. citizens if they qualify and if the organization is registered with the IRS as a tax-exempt nonprofit corporation.

How to Avoid the Pitfalls of Volunteering Abroad

Steering clear of the typical misconceptions of volunteering will help you make the most of your experience. Here are some pointers.

BE REALISTIC

As noble as it is to want to help people, expecting to change the world may be a bit unrealistic. If this will be your first time traveling abroad or visiting a developing country, the impact that the locals make on you may be much bigger than the impact you make on them. Sometimes volunteers have to adjust their expectations and plans once they learn about the realities of the community in which they live. Do not be disillusioned if all of your wonderful plans for making a difference don't move forward in the ways you hoped they would. You probably still affected the people with whom you had contact, but perhaps your influence will only be seen in years to come, like in the inquisitive child you inspired to be the first person in the village to go on to university.

EXPECT TO WORK

Some volunteers are disappointed and slightly put off to discover that the work involved in a volunteer program, like working on a kibbutz in Israel, a farm in Australia, or a school in Guatemala, is *real* work. That means the experience may have some of the same pros and cons as any real-world job, except that you generally will not be rewarded for your hard work with a paycheck. Your reimbursement for volunteering will be mainly in nonmonetary terms—hands-on work experience, training, college credit, tax deductions, possibly room and board, a unique cultural learning opportunity that cannot be found elsewhere, and of course, the satisfaction of helping others and making a difference. If you would rather receive a paycheck for your work abroad, read chapters seven and eight and consider all-expenses-paid, professional volunteer positions, paid internships, teaching, or other paid options. In addition, a few programs that classify themselves as volunteer programs, like the Peace Corps or the MBA Enterprise Corps, will cover your expenses and provide a stipend or per-diem salary. You might also ask about working for a volunteer organization as a paid, in-country group leader.

BE TOLERANT AND FLEXIBLE

Volunteer projects vary widely, from archaeology excavations to ocean research to AIDS services, so an open mind and flexible attitude are essential. Harsh living conditions, combined with the cultural differences that exist between you and your group or the local community can make for some tense situations. It doesn't help anyone when people complain about the hard working conditions, hot weather, lack of hot water, or other discomforts. Although you may feel annoyed from time to time, it is essential to make the best of the situation.

Planning Your Volunteer Adventure

Whether you use the Internet to search for the most desirable volunteer program or contact a specific organization directly, you will need to know the cause you wish to support, the country in which you wish to work, and/or the type of work you would like to do. Volunteers work for diverse causes, from fighting for human rights in China to restoring ancient monuments in Central America. Read through the programs listed in this section to get an idea of the areas in which you can volunteer and the types of exciting and diverse programs that exist. Then, if you want to see more options, take a look at the umbrella Web sites, which have more comprehensive listings. It may be helpful to choose the areas that interest you first and then select the countries where you would like to volunteer. Which of the areas listed below spark your interest?

Agriculture	Disarmament	History	Religion
AIDS Education	Disaster Relief	Human Rights	Renovation
Animal Studies	Earth Science	Homelessness	Scientific Work
Archaeology	Ecological Concerns	Ideologies	Senior Services
Arts	Education	Irrigation	Sex Education
Children and Teens	Environment	Justice	Social Work
Conservation	Ethnic Minorities	Manual Labor	Spirituality
Construction	Festivals	Marine Research	Teaching
Culture/ Anthropology	Gender Issues	Nature Conservation	Third World
Disabled People	Health-Related Issues	Refugees	Other

Most volunteer opportunities overseas are available through exchange organizations, nonprofit groups devoted to a cause, government agencies, or academic institutions. However, if you are keen to work with a particular organization and are willing to donate your time and skills, you may be able to arrange your own volunteer stint. Volunteer projects are located mostly in Asia, Africa, Central and South America, and Eastern Europe. Depending on your special interests, you may be able to arrange work in Western countries as well.

Programs are classified as expeditions, internships, traineeships, or volunteer programs. Be aware that some programs are specific to a profession or age group. Married couples, couples with children, and people with special dietary needs usually are welcome and can be accommodated. Always ask what provisions the program can make for you. Use the contact information to request an application and free catalog or brochure, which are essential for gaining a clearer understanding of each program. Many programs allow you to request information and applications online.

Once you have decided on a program that interests you, apply. Sometimes there is an application or registration fee. Upon acceptance into the program, you will receive an information packet with tips about fund-raising if you have to fund your own volunteer stint. Start planning fund-raising activities as soon as possible. You may also want to apply for scholarships or financial aid, if it is available. Students should make sure they can get college credit if they need it.

Getting Started: Using the Internet to Research

The far-reaching capabilities of the Internet will bring you into contact with a large variety of volunteer organizations worldwide. Many volunteer opportunities are projects coordinated by local governments, ministries, and communities. By using umbrella Web sites, like Idealist.com, to search for volunteer programs in economic development, you might come across the Israeli voluntary humanitarian organization, Aid Without Borders. They offer professional volunteer positions to psychiatrists, social workers, and nurses, among others. You might also find volunteer opportunities to work in graphic design, public speaking, and special-events planning with the International Energy Foundation's regional headquarters around the world. If you do a search by country, you may find that through the Trust of Americas Foundation you can take part in an all-expenses-paid, two-month volunteer program to Central America, the Caribbean, or Ecuador, to teach educational and disability organizations how to use computers and the Internet.

Your Internet search might also locate programs that encompass language learning, internships, and unpaid work that coincide with the offerings from the other chapters in this book. The overlapping nature of each organization's programs means that it is best to refer to the organization's Web site for a complete listing of the volunteer areas. For example, although the Peace Corps is probably best known for its teaching positions, it also offers volunteer work in agriculture, business, and community development. And though Earthwatch, by the nature of its name, might be more often associated with environmental projects, it has interesting volunteer opportunities in the fields of archaeology and health. It is always advisable to view each organization's program and geographical listings or send away for complete information.

Review the umbrella Internet sites listed below to narrow down your choice of organizations.

★ IDEALIST

www.idealist.org

Idealist.com is a project of Action Without Borders and has a thorough database for searching nonprofit organizations, volunteer programs, internships, jobs, services, resources, events, career fairs, and career information. The database's capabilities for locating volunteer opportunities around the world allow you to narrow your search by various criteria, including a particular country, area of focus, and the skills required to take part. You can also browse all volunteer opportunities by country only.

Nonprofit internship and job listings can also be found through the Idealist's Nonprofit Career Center. Individuals can sign up for personalized email messages that include the information you've chosen to receive regarding volunteer and job openings, events, and resources. You can also design the ideal volunteer opportunity according to your interests, skills, and schedule, and then search organizations to find a match. These volunteer programs are open to professionals in many fields, students, retirees, and anyone else interested in lending a helping hand.

★ International Volunteer Programs Association (IVPA)

www.volunteerinternational.org

P.O. Box 381161
Cambridge, MA 02238
Email: Ivpa@volunteerinternational.org

International Volunteer Programs Association is composed of nonprofit, nongovernmental organizations that promote international volunteer and internship exchanges. International Volunteer Programs send individuals or groups abroad periodically and for various lengths of time to work as volunteers, interns, or lay missionaries. You can choose to search the database by world region, country, type of work, or project duration. Additionally, you'll find information about why one should volunteer abroad, how to overcome obstacles, how to find the right volunteer program, what is involved in fundraising, and more.

★ Volunteer Abroad

www.volunteerabroad.com

Volunteer Abroad is an excellent directory of international volunteer programs that seems to target the student market. You begin your search by choosing your desired country, and then sort through the program listings that appear for that country. In the "Why Volunteer Abroad?" section you can determine if volunteering is really for you and find out what you can expect to gain from such an experience. This site is part of the GoAbroad.com network, one of the most comprehensive international education and alternative travel sites on the Internet. It also has related links to study-abroad programs (www.goabroad.com), internships (www.internabroad.com), jobs abroad (www.jobsabroad.com), teaching overseas (www.teachabroad.com), language schools (www.languageschoolsguide.com), adventure travel (www.adventuretravelabroad.com), and other travel information.

Your Volunteering Abroad Options

As mentioned above, the best way to get started is to identify the causes that you feel strongly about and the geographical area where you would like to gain practical experience. If you're not sure about either of these things, you might want to refer back to the self-survey and planning section (see page 60) to determine your personal and professional goals. The areas listed below are a sampling of the fields in which volunteering abroad is most common.

ARCHAEOLOGY

Archaeological fieldwork can be one of the most fascinating ways to learn about ancient cultures. If you are interested in cultural anthropology or archaeology, or are just itching to do something different, consider an archaeological field expedition. If you take part in a program to learn about lost civilizations of the Amazon in Brazil, you might work with scientists, students, and other volunteers in an international research team. You may learn how to collect samples of cultural remains, map locations with a global positioning system, create and maintain survey grids and excavation units, collect botanical specimens for dating, and record and catalog archaeological finds.

Can you rough it in the field? Conditions could consist of basic bathing facilities, cold water, and only occasional electricity. Of course, roughing it is part of the adventure—with riverboats as your only mode of transport, fresh fish and forest fruits as your main food sources, and the wild sounds of the Amazon as your nightly entertainment.

★ **Council on International Educational Exchange (CIEE)**

www.ciee.org

205 East 42nd Street
New York, NY 10017
Toll-free: 888-268-6245

CIEE has a wide variety of offerings, volunteering on archaeological digs being one of them. Select projects can include unearthing Roman fortresses, Celtic burial grounds, or Phoenician villages. The work can be hard labor in hot temperatures, and may include removing dirt from the site and carefully sorting through the remains with small spoons and brushes to uncover pieces of the ancient past. As an archaeological volunteer, you will join a team of ten to twenty volunteers from other countries in two- to four-week projects.

CIEE has over eleven hundred volunteer and work projects in over twenty-five countries. The host community will provide group accommodations that may range from sleeping on a floor mat or in a tent, to staying in a youth hostel or community center. Groups live, work, plan their meals, and travel together. The program is open to U.S. residents at least eighteen years old, though the average age of volunteers is between twenty and twenty-five. No particular skills are required. You'll learn all you need to know once you get there. Prior language experience is not necessary, but it always comes in handy. You'll be responsible for paying for your passport, visa, and travel arrangements, but CIEE can help you arrange what you need. The program fee is about $300 USD for a two- to four-week project. CIEE's other volunteer projects focus on construction, renovation, environmental concerns, and culture.

★ University Research Expeditions Program (UREP)

http://urep.ucdavis.edu

University of California, Davis
1 Shields Avenue
Davis, CA 95616
Phone: 530-757-3529
Fax: 530-757-3537

UREP's motto is "adventure with a purpose." Join University of California scientists on two- to four-week-long research projects investigating critical issues of worldwide environmental, human, and economic importance.

People from all walks of life—from students to professionals to active seniors—can join in the fun. Archaeology, the arts, geology, culture, and conservation are just a few of the areas in which you can volunteer. Study the Bronze Age in Hungary by taking part in an excavation to reconstruct prehistoric domestic life, or learn about the native plants of Indonesia by helping scientists gather information about the roles they play in daily life and ceremonies, and how they are used in local medicines. Groups are between four and eight people. Costs vary per project.

For more research and conservation programs, see the Earthwatch Institute listing (page 199).

BUSINESS AND ECONOMIC DEVELOPMENT

Business professionals and MBA graduates can contribute their knowledge in influential ways to develop foreign businesses, create jobs and better social and economic conditions, and improve health-care facilities. Through two USAID-sponsored organizations, the International Executive Service Corps (IESC) and Citizens Democracy Corps (CDC), business professionals can use their skills to assist business enterprises in developing countries in Central and South America, the Middle East, Asia, Eastern and Central Europe, Russia, and Africa.

Costs for participation are generally covered by the host company, the sponsoring organization, an MBA program, or other sponsors. If your business experience is limited and you do not have an MBA, you still may be able to volunteer in a business setting. For instance, Peace Corps volunteers are more frequently being asked to work on economic development and business-related projects in developing countries (see page xx).

★ Citizens Democracy Corps (CDC)/MBA Enterprise Corps

www.cdc.org

1400 I Street, NW, Suite 1125
Washington, DC 20005

Toll-free: 800-394-1945
Fax: 202-872-0923

The CDC was formed in 1990 to help Central and Eastern Europe and Russia transition to a market economy with the help of American businesspeople. Volunteer advisors, drawn from the top ranks of U.S. businesses, have at least eight years of experience at the senior management level. (MBA students from select schools can participate in the MBA Enterprise Corps, which is discussed below.) Volunteers are generally active managers or owners of their own businesses, and choose to volunteer several months of their time to help and mentor small- and medium-sized businesses in Azerbaijan, the Balkans, Central Asia, Guatemala, Russia, Thailand, and the Ukraine. Business volunteers provide individual management and technical support in projects ranging from modernizing manufacturing processes and financial reporting systems to increasing sales. In addition, volunteers help obtain financing; create new markets for buyers, sellers, and partners in the United States; and bring together regional clients to further education and partnerships. The CDC offers continuing education and involvement that helps promote local, regional, and community business relationships.

If you don't have extensive senior-level business experience, but are in the process of obtaining an MBA, the MBA Enterprise Corps (www.mbaec-cdc.org) is a smart way to help businesses in other countries while putting your theoretical knowledge to use. Applicants must be at least in their second year or a recent graduate of a participating MBA school. During the fifteen-month assignment, students act as management consultants to companies transitioning to a market economy. Job responsibilities may include business-plan writing, financial analysis, product development, joint venture negotiation, and more. Previous international and language experience is desired, but not required. Costs are covered by the host company, which pays for a furnished apartment, utilities, and local expenses; the Corps, which pays for airfare and language training; and the MBA program, which assists in job placement upon your return.

Working in the MBA Enterprise Corps

by Bridget Johns (Bridget_johns@hotmail.com)

My first international work experience was in Poland. In fact, prior to moving to Poland, I had traveled very little and had never lived outside of the United States—I grew up on a farm in rural Pennsylvania. My decision to join the MBA Enterprise Corps was based on two things. First, I had a hunch that I might like to work in an international capacity, but without any international experience, I knew I wasn't going to land a very interesting international job. The Corps seemed to be a fairly low-risk opportunity to gain this experience, as well as learn another language. Second, I had developed an interest in Eastern Europe from a college internship centered around the Bulgarian economy.

As an MBA Enterprise volunteer, I worked for a third-party logistics company in Warsaw. My role was part business development, part troubleshooter, and part internal consultant. Because I only had a one-year contract, my job was very project oriented. I spent a lot of time working on a proposal and then the implementation of a contract to support a major telecommunications company as they participated in the building of the mobile phone system network in Poland. I also worked on a sourcing project for a major restaurant chain, setting up a preretail operation for one of our customers, doing business-plan and financial modeling for a couple of new business ventures, and working on some internal processes. The job was self-motivated and self-directed (as is the case with many international jobs or with loosely defined jobs in fast-growing organizations). My boss realized there was some value in having a newly minted MBA student working for peanuts, but in terms of what the work actually was, I was left mainly to figure it out on my own.

The MBA Enterprise Corps program is the best thing I've done personally or professionally. It provided a fantastic platform for an international career, it allowed me to travel to places I likely would have never seen otherwise, and it gave me a tremendous appreciation for the business environment in the United States, with its modern offices, state-of-the-art computers, copiers, technological infrastructures, and voice mail. Although I had to forgo a traditional salary for over a year, I think this is an amazing program and a great place to start if one is serious about an international career.

I believe I would not be doing what I am doing today—director of business development for a major home-textile manufacturing company—had I not participated in the Corps. Prior to business school, my background consisted mainly of equity research, so it would have been easy to get pigeonholed into a banking career after finishing my degree. The Corps experience made me stand out from my competitors. It also gave me the confidence to work in any business environment. Because of the diverse nature of my position, I learned about many aspects of international business, not just the cultural, financial, and marketing aspects. Not only did I gain valuable work experience, I also made amazing contacts that continue to help me in my job to this day.

Read more about Bridget Johns's international experiences in chapter eight.

THE INTERNATIONAL EXECUTIVE SERVICE CORPS (IESC)

www.iesc.org

333 Ludlow Street
P.O. Box 10005
Stamford, CT 06904
Phone: 203-967-6000
Fax: 203-324-2531

IESC is the world's biggest nonprofit business development organization. Its mission is to help small and medium-size businesses in emerging markets increase their competitiveness. American senior-level business volunteers act as ambassadors of goodwill, sharing skills and knowledge with professional counterparts in more than fifty developing countries. IESC's services include gaining access to global markets, creating strategic alliances, importing and exporting products, obtaining financing, market research studies, business planning, preinvestment screening, and postinvestment assistance. The U.S. Agency for International Development (USAID) is the main funding sponsor of IESC's programs, but corporations, organizations, and individuals contribute as well. IESC has provided results-oriented solutions to businesses as well as nonprofit and government organizations in 120 countries over the last thirty-four years. You can search IESC's database of volunteer opportunities in fifty countries by location, required skills, and program duration. Sample listings include a mergers and acquisitions speaker needed in Panama for one day, or a volunteer with general management expertise needed in Bulgaria for six weeks.

UNITED NATIONS VOLUNTEER PROGRAM (UNV)

www.unv.org

United Nations Volunteers–Headquarters
Postfach 260 111
Bonn, Germany

Phone: 49-228-815-2000
Fax: 49-228-815-2001

If you have at least seven years of solid work experience, you are eligible for the United Nations Volunteer Program (UNV) in international development, projects that involve feasibility studies, employee training, workshop organization, company advising, banking, health and safety support, environmental management, industrial design, and more. Your résumé is placed in a resource bank that can be referred to on short notice. Volunteers provide their skills and knowledge for free, but travel and living expenses are paid.

Also, through the United Nations TOKTEN (Transfer of Knowledge through Expatriate Nationals) program, skilled expatriates from developing countries may return to their home country as a volunteer to work with the government, businesses, and academic institutions for three weeks to three months. See the United Nation's Web site for a complete list of volunteer opportunities.

COMMUNITY DEVELOPMENT

Volunteering to help a community involves a variety of enlightening projects. Organizations like the Council on International Educational Exchange (CIEE) and Volunteers for Peace offer programs that provide particular services to a community, such as helping preserve archaeological treasures, improving quality of life by building community centers and schools, maintaining a region's natural beauty by creating hiking trails, or providing a social service by teaching children. Community development projects occur mostly in developing countries, although a few can be found in Western countries as well. Volunteers usually have the unique opportunity to meet with community leaders in the countries in which they work.

★ Council on International Educational Exchange (CIEE) International Volunteer Project

www.ciee.org

205 East 42nd Street
New York, NY 10017
Toll-free: 888-COUNCIL

The goal of CIEE's International Volunteer Project is to promote international cooperation and understanding. Council sponsors community service and volunteer projects in thirty countries for participants from all nationalities, ages eighteen and up (the average age is twenty to twenty-five). Group sizes average ten to twenty volunteers who come from many different countries. Over six hundred service projects are organized in over thirty countries each summer in just about every area of community service you can think of. There are four basic types of projects offered each year: (1) Archaeology: clear sites, excavate, catalog artifacts; (2) Nature conservation: build hiking paths, plant to control erosion, create garden spaces; (3) Construction and Renovation: restore monuments, refurbish youth centers, build playground areas; and (4) Social Service: organize recreational programs, and care for children, the elderly, or people with special needs.

Projects vary greatly from one country to the next. You could patrol beaches to protect sea turtles along the coast of Mexico or build playground areas in the coastal villages of Turkey. You could develop hiking trails in the breathtaking countryside of Germany's Black Forest or restore rampart walls in the ancient medinas of Morocco. Volunteers live together as a group in a school, community center, hostel, or tent, depending on the project and the host country's resources. Volunteers prepare meals together and organize leisure activities as a group. Programs last two to four weeks, and run mostly from June through September. Program fees are about $300 USD, not including travel costs.

Council also has a Work Abroad program, a Teach in China program, and a variety of summer, semester, or yearlong study programs for students, which can be combined with a volunteer project to extend your stay abroad.

Cross Cultural Solutions

www.crossculturalsolutions.org

47 Potter Avenue
New Rochelle, NY 10801
Toll-free: 800-380-4777
Phone: 914-632-0022
Fax: 914-632-8494

Cross Cultural Solutions is a nonprofit organization dedicated to providing humanitarian assistance to China, Ghana, India, Russia, and Peru. Some people have described its programs as a "mini-stint with the Peace Corps." Each month twelve to eighteen volunteers are sent overseas to work on grassroots projects for two weeks to six months. Volunteers from all over the world work in small groups of two to three people to promote health care, education, and social development. For example, volunteers may teach English to adults and children, teach basic business principles, start programs with local women's groups, assist health-care workers, or provide occupational therapy. All prospective volunteers must fill out a personal data form and a skills and interests survey that helps determine project placement and duties. Fees vary depending on the length of the program and the country. You do not have to pay all the fees by yourself if you do a bit of fund-raising, apply for grants, or get support from local businesses. Contributions are tax deductible. See the Web site for fund-raising tips. Students may be able to receive credit toward their studies, so be sure to ask.

Global Routes

www.globalroutes.org

1814 7th Street, Suite A
Berkeley, CA 94710
Phone: 510-848-4800
Fax: 510-848-4801

Global Routes offers teaching and community service project internships in Asia, Africa, Central America, and United States for students ages seventeen and older. Programs are best suited for people who have a passion for adventure, contribution, cross-cultural immersion, and personal growth. The goal of the program is to bring people with different worldviews together to create a global community. Global Routes interns are assigned to remote villages in pairs, where they'll teach in local schools and complete at least one community service project. Interns generally teach English, math, science, environmental education, or health, and may also choose to coach soccer, volleyball, basketball, or debate; direct plays or choir; teach guitar, martial arts, or a craft; or initiate a dance troupe or poetry club. Each intern lives with a separate family in a simple, traditional home, which gives them a sense of the natural rhythm and flow of daily village life.

In partnership with local leaders, interns assess the needs of the community when designing an appropriate community service project. You may start a small enterprise, form a women's cooperative, deliver health education workshops, or construct a water tank, clinic, or playground. In Ecuador, for example, interns have an in-country briefing where they take personalized Spanish classes and develop teaching skills in addition to learning about the school, rural village life, and environmental issues facing the country. Most Ecuador interns focus their teaching on environmental education and English. Depending on the needs of the village, interns may choose to direct biointensive organic gardening projects, paint educational murals, or start a jewelry business with a women's cooperative.

During the final ten days, interns are debriefed while traveling on a journey of their choosing. Programs last three to four months and are offered year round. A working knowledge of Spanish is required for Ecuador and Costa Rica. Costs are $3,950 to $4,250 USD and scholarship aid, fund-raising options, and college credit are available.

★ Service Civil International–International Voluntary Service (SCI-IVS USA)

www.sci-ivs.org

SCI-IVS USA
814 N.E. 40th Street
Seattle, WA 98105
Phone: 206-545-6585
Fax: 206-545-6585

SCI-IVS M/LTV (Medium/Long-Term Volunteering)
205 North Plain Road
Great Barrington, MA 01230
Phone: 413-528-1307
Fax: 801-906-7716

SCI-IVS organizes international work camps and medium/long-term volunteer projects worldwide. A work camp is a group of two to twenty international volunteers who assist a community with a project for two to three weeks, mostly during the summer months. The medium- to long-term volunteer projects run from one to twelve months year-round. Although the minimum age requirement is eighteen, the average age is twenty-one to twenty-five, and volunteers of all ages are welcome. The goal of the program is to promote international goodwill through friendship and community service.

Work camps provide a unique opportunity for individuals to combine their energies to address issues that are vital to our global future. Subject areas include archaeology; agriculture; construction; the environment; work with children, the elderly, and the handicapped; renovation; the arts; culture; festivals; manual labor; and social work.

Volunteers must be prepared to live and work in a simple, communal environment, preparing meals, working, and relaxing together. The work varies, but is usually physical—building, constructing, and weeding—or social—working with children or taking part in arts-related projects. Volunteers can renovate a school in a mountain village of Greece, create trails at an "Eco Study Park" in Japan, prepare for a human rights festival in Sweden, work on an organic farm in the Swiss Alps, or join an archaeological dig in Russia.

Volunteers often register for multiple work camps in the same or different countries to prolong their trip abroad. For long-term projects, volunteers must have participated in an SCI-IVS work camp or have equivalent experience. Fees range from $195 to $500 USD, not including transportation. College credit for students is available.

CULTURAL PROGRAMS

There is no better way to understand another culture than by living and working among its citizens. Some programs, like the Global Citizens Network, integrate cross-cultural learning elements into its global volunteer programs. Other organizations, like Amizade, offer volunteer opportunities as a unique complement to regional travel. This type of alternative travel gives you a chance to live and interact with locals while helping their community. It is a combination of community service, recreation, and firsthand involvement. So while you offer your medical knowledge, teaching talent, or physical labor to improve a community, you further your own cross-cultural understanding by taking part in activities designed to help you learn about the local culture and customs.

★ AMIZADE

www.amizade.org

Amizade Ltd.
367 S. Graham Street
Pittsburgh, PA 15232
Toll-free: 888-973-4443
Fax: 412-648-1492

Amizade is a nonprofit organization dedicated to promoting cultural awareness through volunteerism and community service. Amizade's original goal was to save the rain forests by helping the people who live there. Nowadays Amizade arranges educational, environmental, health, welfare, and service projects around the world in response to invitations from the communities in need. For example, in 1996, Amizade worked with Pastoral do Menor, a local nonprofit group in Santarem, Brazil, to equip street children with the vocational training they need to develop skills, obtain a job, and have a better life. Amizade organizes volunteer vacations in Brazil, Bolivia, Australia, Yellowstone, and the Navajo nation. The only prerequisite for joining is a willingness to help out where needed.

★ Global Citizens Network (GCN)

www.globalcitizens.org

130 N. Howell Street
St. Paul, MN 55104
Toll-free: 800-644-9292
Phone: 651-644-0960
Fax: 651-644-0960

GCN seeks to create a network of people who are committed to the shared values of peace, justice, tolerance, cross-cultural understanding, and global cooperation; to the preservation of indigenous cultures, traditions, and ecologies; and to the enhancement of the quality of life around the world. Cross-cultural volunteer expeditions are available to people of all ages, including families, in a variety of group sizes.

GCN sends small teams of volunteers to rural communities around the world to immerse themselves in that community's daily life. The teams work on projects initiated by locals, which could range from setting up a library and teaching business skills to building a health clinic or planting trees to reforest a village. A unique and integral component of the experience includes participation in daily cross-cultural learning activities such as visiting a nearby tea factory, learning a local dance, or discovering the rich traditions in the area. Volunteers stay in homes or as a group in a community center. Meals are either eaten with your host family or communally prepared and shared with your project hosts. Projects last from one to three weeks year-round, and cost between $550 and $1,600 USD, not including airfare. Children aged five to twelve travel for half price, and all fees are tax deductible.

★ Kibbutz Program Center

www.kibbutzprogramcenter.org

633 3rd Avenue, 2nd Floor
New York, NY 10017
Toll-free: 800-247-7852
Fax: 212-318-6134

The Kibbutz Program Center offers individuals and couples ages eighteen to twenty-eight a chance to study Hebrew and work on a kibbutz. A kibbutz is a traditional communal society unique to Israel. Although less than 3 percent of Israel's population lives on kibbutzim today, they still exist as a subculture and symbol of the "real" Israel.

Participants live two or four to a basic room that includes meals, sheets, blankets, work clothes, and laundry service. For five months you will study Hebrew for three days a week, and work on the kibbutz to pay your keep for another three days (or do a combination of language and work per day). You can also participate in cultural trips, seminars about current Israeli affairs, and other enrichment activities as part of the program. Work on the kibbutz can vary, but it typically consists of agricultural projects, kitchen duty, working with children, laundry, or factory work. Although you can request the type of work you would like, you will eventually be rotated into different assignments so that everybody has a chance to work at different tasks. Most kibbutzim have pools, tennis courts, recreation rooms, and weekly movies for your free-time enjoyment. Every other weekend you have time to travel on your own. Participants (called *ulpanists*) come from a variety of cultural and ethnic backgrounds, though a large majority are from Eastern Europe, Russia, South America, and Europe. There is a $250 USD registration fee, a $600 USD participant fee, and a $100 USD fee if you want to receive college credit for your participation. You are also responsible for your own airfare, extended insurance, and personal items.

ENVIRONMENTAL PROTECTION AND RESEARCH

Volunteers are needed worldwide to help protect rain forests, oceans, and national parks; to clean up after environmental disasters like oil spills in the ocean; and to research how we can continue to preserve our planet's fragile environment. Volunteers can work on projects in exotic locations, from the Great Barrier Reef in Australia to the Kutai National Forest Reserve in Borneo.

Some organizations, like the Coral Cay Conservation Expedition (see below), work in specific areas, like tropical forests and coral reefs. Others, like Earthwatch Institute (see page 199), research endangered ecosystems, study ocean biodiversity, and work for the survival of black rhinos. Programs like the Global Service Corps can be both experiential and educational in nature, allowing participants to live and work side by side with the locals. If becoming directly involved with environmental protection and research is your goal, you should have no problem finding a volunteer program that needs your skills.

★ CORAL CAY CONSERVATION (CCC) EXPEDITION

www.coralcay.org

154 Clapham Park Road
London SW4 7DE
United Kingdom
Tel: (44) (0) 171 498 6248
Fax: (44) (0) 171 498 8447

Coral Cay offers conservation expeditions in Honduras, Borneo, the Philippines, the Seychelles, the Red Sea, and the Mediterranean for volunteers ages sixteen to seventy. The goal of the program is to help alleviate poverty by providing countries with the information they need to protect and sustain their tropical forests and coral reefs. Therefore, volunteers gather information that can be used to establish management schemes to reach this goal. Programs vary in duration from two to twelve weeks, and training is provided so all volunteers make an effective contribution to the project.

For example, the recent collapse of Indonesia's economy, combined with a population fast approaching two hundred million and poorly regulated fisheries and forestry practices, has led to widespread destruction of forests and coastal habitats throughout Borneo. In 1998 a team of CCC volunteers joined forces with the Tanjung Bara Dive Club to survey and monitor reefs adjacent to the Kutai National Forest Reserve. The success of this project led to the request for further collaboration with CCC, and in 1999, teams of CCC volunteers joined local counterparts to continue and expand surveys of the forests and reefs along the East Kalimantan coastline.

Program fees, which may be tax deductible, are 450–2,550 British pounds ($310–1,767 USD), not including transportation.

★ **Earthwatch Institute**

www.earthwatch.org

680 Mt. Auburn Street, Box 9104
Watertown, MA 02471-9104
Toll-free: 800-776-0188
Phone: 617-926-8200
Fax: 617-926-8532

Earthwatch supports scientific research and promotes global citizenship and education for a sustainable planet. Earthwatch is a worldwide scientific research expedition for people ages sixteen to eighty-five who are driven by a desire to know how the world works and how they can help make it work better. Earthwatch members come from forty-six countries and work in groups of eight to fifteen to gain hands-on scientific field-research experience working with the top scientists who lead the teams. Members either participate in research or conservation projects. Areas include endangered ecosystems, ocean biodiversity, global change, cultural diversity, world health, and archaeology. Basic research expeditions involve understanding how the Earth and its systems work and providing the raw data that other researchers use to focus their applied work. Conservation programs address immediate problems like saving black rhinos and rescuing disappearing cultures. Earthwatch's conservation programs have produced new laws, national parks, museums, and ways of feeding and caring for the world's disadvantaged people.

The accommodations, food, and required level of fitness vary, but there is an appropriate program for every taste and comfort level, be it camping or a luxury hotel. Projects can run from one week to one month, volunteers receive training, and English is spoken. Costs range from $600 to $4,000 USD plus travel costs, and are tax deductible. Competitive grants are available for students and teachers, and subsidies and fellowships exist through corporate and educational sponsors. You can search the Earthwatch database by region, month, or subject to find the best expedition to meet your needs.

★ Global Service Corps/Earth Island Institute

www.globalservicecorps.org

300 Broadway, Suite 28
San Francisco, CA 94133
Phone: 415-788-3666
Fax: 415-788-7324

In the Global Service Corps (GSC), you work on community service projects and live with local residents while learning about the host country.

Volunteers must be undergraduate or graduate students, twenty years or older, and projects are organized in Kenya and Thailand.

The goal of the program is to improve the well-being of the Earth and its inhabitants. GSC's short- and long-term projects target three primary areas: Environment (Global CPR—Conservation, Preservation, and Restoration), Health (public health and clinical care), and Education (training seminars and classroom teaching).

Short-term programs (two to four weeks in duration) are experiential and educational. Long-term programs (two to six months in duration) are geared to independent and self-motivated students who have specific skills to share and interests to pursue. Participation in a long-term project is an excellent way for students to fulfill internship or independent study requirements.

Short-term trips serve as orientation for the long-term programs. In GSC's long programs your interests and abilities will be matched to the needs of the partner organizations. Projects might include assisting with conversational English training, or working on environmental projects, community health care, or HIV/AIDS awareness activities. The short-term program fees, which are tax deductible, are $1,695 to $1,795 USD, plus airfare, and the long-term program, which is also tax deductible, costs $2,520 to $2,820 USD, plus airfare. Optional college credit is available, so be sure to ask.

For other environmental volunteer opportunities, check out these programs: Global Citizens Network (see page 196) and the University Research Expeditions Program (see page 184).

HEALTH PROGRAMS

If you work in the health or medical fields, you can use your skills to train professionals in Bangladesh, Brazil, Jamaica, or Vietnam, and provide treatment, continuing education, lectures, and workshops in a variety of volunteer programs organized by health-related organizations. Whole populations in developing countries are threatened by the lack of basic health care, the rising costs of medicine, the unavailability of government funding, the rise in AIDS/HIV-infected people, and the gap in care for women and children. You can make a difference by volunteering your services, whether you are a health professional or not, to communities in need.

★ Doctors Without Borders/Médecins Sans Frontières (MSF)

www.doctorswithoutborders.org

6 E. 39th Street, 8th Floor
New York, NY 10016
Phone: 212-679-6800
Fax: 212-679-7016

Doctors Without Borders (MSF) delivers emergency medical care to communities and victims of armed conflict, epidemics, natural disasters, and others who lack decent health care due to geographic or social isolations. Doctors, midwives, mental health specialists, pharmacists, and humanitarian officers are just a few of the types of medical and nonmedical professionals needed at MSF. Although MSF sends over three thousand volunteers to ninety countries each year, they have an ongoing need for professionals and are always trying to increase their volunteer pool. Volunteers have to interview at one of the MSF offices and commit to at least a six-month assignment, although nine- to twelve-month assignments are more likely. Costs are covered by the organization and a monthly stipend is provided.

★ Health Volunteers Overseas (HVO)

www.hvousa.org

P.O. Box 65157
Washington, DC 20035
Phone: 202-296-0928
Fax: 202-296-8018

HVO organizes over forty educational projects in twenty developing countries with the goal of improving the quality of the local health care. HVO volunteers must be qualified and experienced professionals and retirees.

You may train local health-care providers in the fields of anesthesiology, dentistry, internal medicine, oral and maxillofacial surgery, orthopedics, pediatrics, and physical therapy, or lecture, conduct ward rounds, and demonstrate techniques in classrooms, clinics, and operating rooms. Program durations vary from several weeks to several months. Family members may accompany the volunteer, and if qualified, may also volunteer. You are responsible for your transportation to and from the site; room, board, and daily transportation are provided upon arrival.

★ Peacework's Medical Volunteer Program

www.peacework.org

305 Washington Street, SW
Blacksburg, VA 24060
Phone: 540-953-1376
Fax: 540-552-0119

Peacework's Medical Volunteer Program arranges international volunteer projects focused on medicine, community health, and nutrition. Volunteers may collaborate with local health and medical services or private international health agencies, or participate in general health and outreach services in urban and rural areas. For example, responding to the devastation caused by Hurricane Mitch in Honduras, Peacework sent health-care volunteers to be involved with intervention services, health evaluations, health education, basic treatments in rural areas, and short-term medical relief in the countryside.

Peacework provides all the services needed by health or medical groups volunteering overseas. This includes international travel, medical certification (if needed), visas, liaison between sponsor and host, room, board, in-country transportation, organization of field and cultural excursions, insurance, and predeparture information. The sponsoring organization or group is responsible for orientation, supplying medical documentation or certification for participants, and on-site supervision.

In addition to medical volunteering, Peacework brings together U.S. colleges and universities with organizations overseas to create service-learning programs that complement a participant's course of study. They collaborate with church groups and nongovernmental civic and service organizations to promote local education, relief, and development efforts. Volunteers for all of Peacework's projects range in age from teens to seniors from all walks of life. Programs are usually one to four weeks in duration, although the length depends on the sponsoring group. Costs vary per program and duration.

The Peace Corps (see page 209) also offers volunteer positions in the health field.

HUMANITARIAN RELIEF AND SERVICE PROGRAMS

By providing humanitarian relief or community service, volunteers can help people and societies at the grassroots level. As a volunteer you'll help refugees of war rebuild their lives, assist in building houses, or teach a community how to grow its own food.

Organizations like Concern America, the United Nations, and Habitat for Humanity look for volunteers whose backgrounds range from basic teaching and construction skills to more specialized skills in the fields of technology and medicine.

★ Concern America

www.concernamerica.org

2020 N. Broadway, 3rd Floor
Santa Ana, CA 92701
Toll-Free: 800-266-2376
Phone: 714-953-8575
Fax: 714-953-1242

Through Concern America, professionals in the fields of medicine, health, nutrition, health education, adult literacy, sanitation, agroforestry, technology, and community organization can volunteer to help refugees in Latin America and Africa. Volunteers do not receive a salary. Instead, Concern America provides round-trip transportation, room and board, an annual trip home, health insurance, a small monthly stipend, support services, and a small repatriation allowance. Volunteers must make a one- to two-year commitment to serve. The minimum age requirement is twenty-one, and knowledge of or a willingness to learn Spanish or Portuguese is required.

★ Habitat for Humanity International (HFHI)

www.habitat.org

Partner Service Center
121 Habitat Street
Americus, GA 31709
Phone: 229-924-6935, Ext. 2551 or 2552
Fax: 229-924-6935

Habitat for Humanity is a Christian organization with affiliates in over sixty countries devoted to providing decent, affordable housing for families worldwide. Habitat for Humanity's Global Village Program is a series of one- to three-week mission trips designed to provide an educational and spiritual experience within a cross-cultural environment, in locations such as Nepal and New Zealand.

By living and working with a host community, Habitat participants have an opportunity to personally witness and contribute to HFHI's worldwide projects. As you learn about poverty housing, international economics, and world affairs, you immerse yourself in the host community's culture, language, and social environment.

Costs vary between $1,200 and 4,000 USD, depending on the country visited, the HFHI affiliate, and the length of the program. Prices include travel and living costs as well as a donation to the host affiliate organization. Many participants engage in fund-raising to earn all or part of their program costs. (See more about fund-raising in chapter nine.) To apply, contact your local HFHI affiliate, which can be located by using the search engine on HFHI's Web site or by calling them direct.

★ Heifer Project International (HPI)

www.heifer.org

P.O. Box 8058
Little Rock, AR 72203
Toll-free: 800-422-0474
Phone: 501-907-2600
Fax: 501-907-2600

Heifer Project International is dedicated to ending world hunger and poverty by providing families with the animals they need to become self-reliant. Children receive nutritious milk or eggs; families earn income by selling the dairy products for school, health care, or better housing; and farmers learn sustainable and environmentally sound farming techniques. Through HPI you can partake in a study tour or work-study program abroad and share your experiences upon returning home. These are one- to two-week trips with groups of ten to fifteen people who are interested in partnering with the world's poor while learning more about the effects of poverty. Projects are led by HPI staff and volunteers. Prices range from a few hundred to a few thousand dollars, depending on the country visited.

★ InterAction: American Council for Voluntary International Action

www.interaction.org

200 Park Avenue South
New York, NY 10003
Phone: 212-777-8210

1815 H Street NW
Washington, DC 20006
Phone: 202-822-8429

InterAction is the United States' leading advocate for humanitarian aid. Consisting of a coalition of over 165 disaster relief, development, environmental, and refugee agencies, InterAction provides free information and advice on finding employment with nonprofit organizations, and for a fee, job resources for working with nonprofit international organizations. The goals of the member organizations are to promote economic development, improve health and education, offer assistance to war victims and refugees, protect the environment, and advocate public policy.

★ United Nations Volunteers

www.unv.org/unvols

United Nations Volunteers–Headquarters
Postfach 260 111
Bonn, Germany
Phone: (49) 228-815-2000
Fax: (49) 228-815-2001

The United Nations accepts qualified volunteers from many countries to work in the following areas: technical cooperation for development, community projects, humanitarian relief and rehabilitation, electoral rights, human rights, and peace-building initiatives. Volunteers should have at least a bachelor's degree, several years of relevant work experience, and a willingness to live and work in challenging conditions. Many professions are in demand, such as computer hardware specialists, environment lawyers, HIV/AIDS home-care specialists, planners, psychologists, accountants, and waste management specialists.

★ Volunteers for Peace, Inc. (VFP) International Workcamps

www.vfp.org

1034 Tiffany Road
Belmont, VT 05730-0202
Phone: 802-259-2759
Fax: 802-259-2922

VFP organizes work-camp programs linking volunteers to work-camp hosts in seventy countries in an effort to promote intercultural education and community service. A work camp consists of twelve to twenty international volunteers who assist a community with a specific project designed to improve the lives of the locals in a tangible way. As part of the work-camp experience, you'll learn about the social, cultural, and political conditions that exist in other countries, discuss issues of common concern with other volunteers from diverse cultural backgrounds, and take excursions to places of interest.

Volunteers must be prepared to live and work in a communal environment, preparing meals, working, and relaxing together. Volunteers, who are generally age eighteen and up (average age is twenty-one to twenty-five), usually work about thirty hours per week. The program costs are incredibly reasonable. The registration fee is $200 USD per work camp, plus a mandatory VFP membership of $20 USD. The registration fee covers in-country expenses, including meals and accommodation for one camp. Because housing in host countries is often simple and free of charge and volunteers prepare their own meals, the cost to live in a foreign country is kept at a minimum. However, your costs of getting to the host country and your spending money should be figured into your budget.

To sign up for a work camp, become a member of VFP to receive VFP's International Workcamp Directory. From this directory of worldwide work camps you simply choose the work camp(s) that interest you, fax the registration form, and send payment to VFP. Placements are usually made within five business days after your payment is received.

TEACHING ENGLISH

Your ability to speak, read, and write in English is a skill in high demand throughout the world. Although English is considered the language of international business, only one-tenth of the world's population can speak it well. By teaching English to students and adults throughout the world, you give them a chance to get an education, find better jobs, and hopefully prepare for a more prosperous future. In addition, by living in the local community and teaching at a local school, company, or organization, you gain a better understanding of the culture, language, social customs, and ways of doing business. It also allows you to make future business and job contacts. Many exchange organizations, like CIEE (see page 127), offer the opportunity to teach English overseas. Also see chapter seven for more programs to teach English abroad.

★ Peace Corps

www.peacecorps.gov

1111 20th Street, NW
Washington, DC 20526
Phone: 800-424-8580

The Peace Corps offers one of the best overseas experiences available. Teaching English is the Peace Corps's most popular program, but it is by no means the only one. Two-year volunteer opportunities are also available in business, environment, health, agriculture, and community development. The Peace Corps, whose motto is "The toughest job you'll ever love," allows you to work at the grassroots level to improve people's lives.

When reviewing applications, Peace Corps administrators consider the whole person—your educational background, work experience, community involvement, and commitment to being a Peace Corps volunteer. Volunteers range in age from eighteen to eighty-six and include married couples (but not their dependents). Peace Corps volunteers speak more than 180 languages and dialects, but previous knowledge is not required as you will receive intensive language training prior to your departure. Although you can request to be sent to specific countries, the matches are made based on where your skills are needed the most. A monthly stipend to cover food, housing expenses, and local transport is provided.

One of the many benefits of joining the Peace Corps is that you may be able to defer payment on certain educational loans during your Peace Corps service. Volunteers with Perkins loans receive a 15 percent cancellation of their outstanding balance for each year of their two-year service. The Peace Corps also partners with graduate schools to offer students the opportunity to combine a graduate degree with Peace Corps experience. Through the Master's International Program, students typically spend one to two years completing course work on campus and then twenty-seven months serving as a Peace Corps volunteer. Your Peace Corp assignment will be the basis for an academic project to be completed upon your return from the Peace Corps. After the completion of your service, you will receive an additional $225 USD for each month you served. Assuming you completed the full two-year service, you will receive $6,075 USD. During every month of service, you will also receive two days of vacation time that you can use to travel at your discretion or to receive visitors.

The Peace Corps's Returned Volunteer Services (RVS) provides returning volunteers with career and educational assistance. As a returned volunteer, you have noncompetitive eligibility status for appointments to U.S. government executive branch agencies for a period of up to one year after completion of your Peace Corps services. That means you will not have to compete with the general public for some federal government positions, as certain federal agencies automatically appoint returned Peace Corp volunteers to fill these openings. (For more details, contact the Peace Corps.)

The Peace Corps experience is one of the best ways to get field experience, especially if you want a career in development work. On the Peace Corps' Web site you can browse the volunteer assignments, read about past volunteers' experiences, find out about Peace Corps events, and apply for a volunteer assignment online. Call the 800 number listed to get contact information for your local Peace Corps office.

★ VIA/Asia Public Service and Educational Exchange Programs

www.volasia.org

P.O. Box 20266
Stanford, CA 94309
Phone: 650-723-3228
Fax: 650-723-3228

VIA (formerly Volunteers in Asia) sends thirty to forty English teachers, ages eighteen to eighty, to Indonesia, Laos, Vietnam, and China every year. As an integrated member of the Asian community, you empower people to communicate in an increasingly global world while at the same time build bridges across the two cultures. You may teach English in a classroom or be an English "resource," providing English-language assistance to nongovernmental organizations. One-year and two-year programs are open to anyone who speaks English with native fluency and has a bachelor's degree. Volunteers must participate in all training sessions, which include a six-day cross-cultural training course at Stanford University, an introductory TEFL (Teaching English as a Foreign Language) course with practicum, language training, and application and selection process. For most posts there are a variety of housing options available, from staying with a family to living among students in university housing.

The two-year program costs $950 USD, the one-year program is $1,975 USD, and the six-week summer program (available to continuing undergraduates from San Francisco Bay Area schools only) costs $1,975 USD. Participation fees cover international airfare, basic health insurance, cross-cultural training, visas, and in-country support. However, be aware of the hidden costs of immunizations, extra health coverage, an introductory TEFL course, and language study. Limited VIA scholarships are available. It is expected that volunteers will raise funds and apply for other scholarships for which they qualify. One- and two-year volunteers are provided a monthly stipend that allows them to live comfortably at the local level.

WorldTeach

www.worldteach.org

c/o Center for International Development
Harvard University
79 John F. Kennedy Street
Cambridge, MA 02138
Toll-free: 800-4-TEACH-0
Phone: 617-495-5527
Fax: 617-495-1599

At WorldTeach, you can make a valuable contribution to communities throughout Asia, Latin America, Eastern Europe, and Africa by teaching a life skill that will open doors of opportunity. For summer teaching programs, you must be eighteen years old to apply. For six-month or year-long programs, you must have completed a bachelor's degree by the start of the program (except for the China program, where you just need to be eighteen or older). It is not necessary to have formal teaching qualifications to apply, however twenty-five hours of teaching ESL or EFL in your community will be required before you leave. No knowledge of the local language is needed as it will be taught during your in-country orientation. Depending on the country, you may live with a host family, in shared housing, or in your own apartment, and you may have to forgo the conveniences of a private bath and hot running water.

The program cost (or volunteer's contribution) of $3,990 USD (summer) and $5,990 USD (full-year) covers international airfare, health insurance, organizational support, field support, and orientation. The good news is that most volunteers raise a large portion, if not all, of the funds by fundraising (see chapter nine). Your (or a family member's) employer may offer tax-deductible matching funds for donations. If you have the right qualifications, you might be eligible for a WorldTeach scholarship. Your host school, community, or government will pay you a small stipend, approximately equal to the salary of a local teacher, to cover your daily living expenses.

VOLUNTEERING ONLINE

Believe it or not, you can volunteer for projects online, from translating brochures to developing Web sites. This is a really great way to get involved with international development projects, lend a helping hand, and use your skills. It usually only requires one to five hours per week, plus you can work on projects when it's convenient—nights, weekends, holidays.

★ NETAID.ORG FOUNDATION

www.netaid.org

336 E. 45th Street, 2nd Floor
New York, NY 10017
Phone: 202-906-6868
Fax: 202-906-6868

Netaid.org and the United Nations Volunteers Program are working together to allow you to volunteer online. You can search Netaid.org's database of volunteer opportunities by professional skills, language skills, topics of interest, and regional preferences. Volunteer opportunities include providing technical assistance and data analysis, computer programming, teaching online, or translating.

★ VOLUNTEERS IN TECHNICAL ASSISTANCE (VITA)

www.vita.org

Suite 710
1600 Wilson Blvd.
Arlington, VA 22209
Tel: 703-276-1800
Fax: 703-276-1800

VITA empowers people, communities, and institutions in developing countries like Guinea, Moldova-Ukraine, and Morocco by providing access to communications, information, and technology. VITA supports entrepreneurs in particular and facilitates information exchange between individuals and organizations. Volunteers provide technical assistance from their own home or office, but are not sent overseas.

Volunteers at VITA respond to requests from the developing world only. Registered volunteers, who are professionals in areas such as agriculture, business, health and medicine, housing, transportation, and sanitation, are placed on an electronic mailing list where incoming requests are distributed. Volunteers with competency in the specific technology area will self-select those questions to respond to.

Requests can vary considerably: teachers in Paraguay are looking for technologies to recycle wastepaper; a small community organization in Namibia requests assistance to teach their indigenous people how to use the forest resources to generate income; a group of blind people in Kenya wish to sustain themselves by learning the technical skills to make soap and soap products. Volunteers communicate via email to answer requests like these.

VOLUNTEERING BOOKS

The number of volunteer opportunities available these days is far more extensive than it's ever been. To help you narrow down the plethora of volunteer programs, here are a few other resources.

★ **ALTERNATIVES TO THE PEACE CORPS: A DIRECTORY OF THIRD WORLD & U.S. VOLUNTEER OPPORTUNITIES**
9th edition, edited by Joan Powell (Food First Books, 2000)

In this handy little guide, you'll gain insight into the lessons learned from the Peace Corps and how to become an informed and aware volunteer.

★ **THE BACK DOOR GUIDE TO SHORT-TERM JOB ADVENTURES: INTERNSHIPS, EXTRAORDINARY EXPERIENCES, SEASONAL JOBS, VOLUNTEERING, WORK ABROAD**
by Michael Landes (Ten Speed Press, 2000)

This is a motivating guide that takes a self-enrichment and holistic approach to finding perfect short-term work or volunteer opportunities. In the "Abroad Adventures" chapter there are excellent descriptions of a wide variety of short-term jobs, internships, and volunteer programs abroad, and how to go about applying for them. Also see the Web site at www.backdoorjobs.com.

★ **The International Directory of Volunteer Work**
7th edition by Louise Whetter and Victoria Pybus (Vacation Work, 2000)

Here you'll be able to search paid and unpaid volunteer positions sorted by country and by the type of work you wish to do.

★ **So, You Want to Join the Peace Corps . . . What to Know Before You Go**
Dillon Banerjee (Ten Speed Press, 2000)

If you are thinking about joining the Peace Corps, this book, written by a former Peace Corps volunteer, answers many of your questions and concerns.

★ **Volunteer Vacations: Short-Term Adventures That Will Benefit You and Others**
7th Edition, by Bill McMillon and Edward Asner (Chicago Review Press, 1999)

This book covers about 250 volunteer organizations worldwide that you can search by cost, length, location, season, or program type.

CHAPTER SEVEN

Working Abroad: Gaining International Experience through Work-Abroad Programs

When fate closes a door, go in through a window.
UNKNOWN

It seems like everyone wants to work overseas these days. Working abroad conjures different images for everyone—a summer bartending job at an Irish pub, teaching English to schoolchildren in a Guatemalan village, taking part in a management trainee program in Frankfurt, or opening a corporate office in Russia. There are many ways to get a job in a far-off destination. Take part in an exchange program to intern, volunteer, or work abroad; work for an international company in their overseas office; or go to your dream destination and create your own exciting work adventure. I know many people who have found or created professional opportunities abroad and they are not any different from you and me. Some have degrees from top international business schools and some don't have degrees at all. Some had role models or relatives with international aspirations, while others had no connections whatsoever. Many are career oriented, and others want to devote themselves to learning a new language or enjoying life in another culture. What they all have in common is an ambition and desire to enrich their personal or professional lives with the rewards of living and working abroad.

If you are a student and want international skills and experience to increase your chances of being accepted into graduate school or to generate better job offers, most universities offer summer work, internships, and volunteer programs through their career development and study-abroad offices or international exchange organizations. If you are a recent graduate or professional, you may be more interested in pursuing an international lifestyle. This chapter will show students and professionals how to easily gain work experience overseas through international work-exchange programs. Then chapter eight will help you take your experiences to the next level by developing professional skills and a global career.

International Job Market Realities

There are two unavoidable catch-22 situations that make getting an international job challenging for many Americans. However, you *can* be successful if you know ahead of time what to expect.

CATCH-22 #1

You need international experience to get hired, but it's difficult to get international experience if no one will hire you.

It's true. In many cases you need some international experience to get hired into a global position. The exceptions are jobs in fast-growing industries, telecommunications being the most recent example, that can't hire consultants quickly enough. If your skills are in high demand in the worldwide job market, then you'll probably have an easier time getting hired by an international company regardless of your experience. Exceptions aside, if you want to work in an international company that will send you overseas, you increase your chances of success by having some overseas experience.

Students and young professionals have many opportunities to do internships, volunteer abroad, participate in training programs, teach, or do other types of short-term work. And that's exactly what this chapter is about: showing you how to get basic experience to put you in a better position for a job overseas. Professionals or career changers who have been out of school for a while will find this catch-22 situation particularly difficult, especially if internships, volunteer programs, or teaching stints abroad are simply not options because you need to keep earning a salary to support yourself and your family. To break into the international world of work, you'll have to assess your skills and figure out how you can apply them in a global setting. You do have an advantage over students in that you actually have work experience in a particular field. This is a huge plus, as professional abilities make you more marketable. If you are serious about going global, "internationalize" yourself by taking a language course, getting involved with the international community, and conducting informational interviews in your industry to find out how to break into the global market. You will find work-abroad options and tips for professionals in the next chapter.

If you've decided to just pack up and go, figuring you'll find a job once you get overseas, consider the second unfortunate reality.

CATCH-22 #2

To work in a company overseas, you need a work permit to get hired. But to get a work permit, you first need to prove you have a job.

Yes, this is officially true in many countries, especially in Europe. To hire you, companies must prove that you have special skills or that you are better qualified than a local would be. In an effort to keep foreign workers to a minimum, many of the Western countries that appeal to foreigners enforce strict work permit regulations. In an international job market where citizens of many other countries are highly educated, speak one or more foreign languages fluently, have global experience, and don't need a work permit (such as in the European Union), it is very difficult for an employer to prove that they need you.

Luckily, there are some solutions. The obvious way around this problem, as discussed above, is to work for a company that will send you abroad and take care of all that nasty bureaucratic stuff for you. These types of positions are generally offered to mid-level and higher management types, IT (information technology) or telecom wizards, and people with many years of international experience. If you don't fall into one of these categories, you're facing catch-22 #1 again. If you can't find a company that is willing to send you abroad, here are a few alternatives to finding or creating international work.

- Arrange a Work Permit through an Exchange Organization

  The easiest way for students, grads, and young professionals to get around the work permit issue, while simultaneously gaining that coveted first experience overseas, is to work, intern, volunteer, or teach through an organized work-exchange program like the ones you'll read about later in this chapter. Exchange organizations were created to facilitate academic and work exchanges across borders. Check out the following programs for more information: Council on International Educational Exchange (see page 232), BUNAC—British Universities North America Club (see page 231), and the Carl Duisberg Society (see page 234).

  If you're a working professional who wants more than entry-level experience—don't despair. There are also a few exchange organizations that focus on professional career development and place people with work experience in comparable paid positions abroad. For instance, The Association for International Practical Training (see page 233) helps professionals (up to the age of thirty-five and in a variety of fields from business and engineering to the culinary arts and architecture), find a placement and get a work permit for assignments of up to eighteen months. You might also be able to arrange a technical or professional exchange with people from other countries who work in your profession. Check out the People to People's Ambassador Programs (see page 168). You'll read more about all of the work-exchange options later in this chapter.

- Obtain a Work Permit for Your Special Skills

  If a country needs professionals with specific skills, they may offer special work permits to people with those qualifications. For example, in the spring of 2000, Germany issued work permits for up to twenty thousand information technology and communications experts from non–European Union countries because the overwhelming need for IT specialists in Germany far outweighed the number of qualified German and European Union applicants. These types of work permits are good for at least five years and allow employees to change companies if desired. (For the New Green Card Regulation in Germany, see this link in English: www.arbeitsamt.de/hst/international/egcindex.html)

  Other countries may issue similar permits for a variety of positions if there is a lack of local talent. In Australia, work permits are assigned according to a point system, with more points being awarded to foreigners who meet particular requirements. For example, if you have a job offer from a company of ten or more people that has been in business for at least two years, you'll receive extra points, putting you in a very favorable position for getting a work permit. You also get more points for having a desired skill set. These skilled workers are in high demand in Australia: IT managers, electrical engineers, accountants, nurses, radiation therapists, and hairdressers. Click to the list at www.immi.gov.au/allforms/modl.htm, provided by the Australian Department of Immigration and Multicultural Affairs for more information about working in Australia.

  There may also be other ways to obtain work permits through the officially recognized organizations that issue them. It is worth contacting foreigners working in your target destination to get the inside story.

- Become a Self-Employed Consultant

  When I lived in Poland, some foreigners worked around the work-permit problem by establishing their own companies and offering their services on a consulting basis. Even if you have never considered setting up your own business, this tactic can be much less complicated than it sounds. Each country has different regulations, so you will have to check into this in detail for your particular country. Give self-employment more than just a passing thought if you are hung up on the work-permit issue. In fact, thinking about what skills or services you have to offer other companies is more or less job hunting in disguise. Repackaging your skills and knowledge as a service with a catchy company name can give you the legal means to work in some countries, at least for the short term. See more about creating your own opportunity abroad in chapter eight.

- Do Freelance Work on an Extended Residency Visa
  An American travel writer I know lives in Germany and works out of her home in Frankfurt. Because she can prove that she has an income and therefore won't be a burden on the German social system, she was able to get a one-year residency permit, instead of the typical three-month travel visa, to live and do her American freelance work in Germany. Because her profession (travel writing in English) does not take jobs away from qualified Germans, she does not need a typical work permit. Depending on the duration and renewal terms of your residency visa, you might also be able to lead tour groups, translate and edit publications, or arrange seminars for the local expatriate community. Each country's rules about self-employment and residency visas will vary, so a quick call to the consulate or embassy is necessary before you move.

- Work in the "Gray" Market
  If you are already living in a foreign country and are looking for part-time or seasonal work in restaurants, pubs, or the hospitality industry, you may fit into the "gray" market. Although it is officially illegal to hold service jobs without a work permit, thousands of people do it everyday in cities from London to New York. It's likely that you would have to leave the country every three to six months to renew your tourist visa, so be aware that the authorities may catch on to your work activities if they see a pattern of entries and exits. Some weigh the risks and figure they can get away with not having a work permit for a few months. The longer you work in the gray market, the larger the risk of getting caught. If you decide to go this route, remember that working without a permit is illegal

and you can get into a lot of trouble, to the point of being banned from visiting the country again.

Rebecca Falkoff (see her "Global Citizen's Perspective" on page 159) gave informational walks at the Colosseum in Rome, in a type of working arrangement that is popular among English-speaking foreigners, albeit a bit vague on the legal side. "We did not actually do tours, we did informational walks; we did not work, we volunteered; we were not paid, we were appreciated. Basically any native English speaker can show up, get a script, and start giving tours. You get paid on commission and tips, so it costs the cultural associations, like Walks of Rome and Discover Roma, nothing. At the start of the summer my tours were terrible, by the end of the summer I was dreaming about the Colosseum every night and spending my weekends studying Roman history. Incidentally, you make piles and piles of money working as a group leader giving informational walks at the Colosseum—more than bankers and lawyers in Rome. The only problem is that the work is seasonal and very unreliable. The walks often get broken up by the police, who are required to respond to complaints made by licensed tour guides."

- Work According to Locally Accepted Ways of Doing Business
  In some countries there may be locally accepted ways of doing business that will allow you to work without an official permit. When Poland was transitioning to a market economy, the immense bureaucratic structures made getting work permits a nightmare for foreigners. Despite filling out the proper paperwork, some people I know worked for several years without ever seeing their work permits. This backup can happen in Spain, Portugal, and many other countries. Working with only a tourist visa while waiting for the official permit to be processed was an accepted way of doing business at that particular point in time, not only in Poland, but also in the neighboring countries of Hungary and the Czech Republic. If the country you want to work in is in a state of transition or rumored to be lax about tracking workers with hazy employment situations, you may have more flexibility. However, make sure

you understand the risks involved. Breaking other countries' laws can get you into more trouble than it's worth.

- Start Your Own Business or Franchise

  If you are bringing money and resources into a country, chances are you will be allowed to officially stay and work. And, many companies will welcome the foreign investment and opportunity to employ locals. As with all major business transactions there will likely be a pile of complex forms to fill out in the local language, so hire experienced legal counsel to help you make sense of the local regulations.

  According to Eve Berton (read her "Global Citizen's Perspective" on page 284), "In Poland, an expat who wants to be self-employed can start a business and be legal without a work permit. Starting a business is a rather complex procedure. It requires a visit to a lawyer to initiate court registration documents, a notary to verify documents, visits to two different tax offices to obtain two different tax identification numbers, long waits, lots of questions (not in English, *never* in English), a bank account, and an accountant to file monthly tax statements. It is an expensive and complex rigamarole (expect a minimum investment of $2,000 USD). Do your research before embarking on this venture. I don't advise anyone to attempt this process without a native helper, whether it's your business partner or a friend. The bureaucratic procedures are very tricky."

- Get a Second Passport

  If one of your parents or grandparents were born outside of your home country, you might be able to obtain a second passport from that country, making it significantly easier for you to work abroad legally. For example, if you are blessed with the luck of the Irish and have a grandparent, or in some cases a great-grandparent, who was born in Ireland, you could have your European Union (EU) passport in hand within three to nine months. Although it can be a hassle to complete all the necessary paperwork, the benefits of having the legal right to work in multiple countries within the EU is worth it. Italy, Germany, and Israel have similar laws. If one of your parents or grandparents come from abroad, it is worth a call to the consulate to find out if you are eligible for a passport from that country.

These are just a few of the ways determined job hunters have been able to keep themselves employed while living abroad. You may find additional loopholes that work to your advantage. Get connected to the expatriate community and find out how other foreigners have found employment.

What Is a Work-Abroad Program?

One of the easiest ways to get a work permit and find a job abroad is through a work-abroad program arranged through a university or an exchange organization. Work-abroad programs are geared mostly toward undergrad and graduate students, recent graduates, and young professionals; however, there are also many overseas teaching and volunteer programs for people of all ages. A work program or work exchange prearranges or helps you secure most (or all) of the following so you don't have to: work and residency permits, accommodations, language courses, and an internship or job. An initial placement or application fee may be charged, but the cost is usually offset by the money you make from your job or by the university credit you receive. Organized programs run for a set period of time, such as the summer, a semester, an academic year, or longer.

The Benefits of Working Abroad

Arranging an overseas job through an organized work or exchange program offers students and young professionals with limited experience many advantages over the traditional route of sending résumés to international companies and hoping for a dream position overseas.

GET YOUR FEET WET THE "EASY" WAY

Get a taste of a foreign culture, language, and work environment with the short-term support of your work program, without the risk of a long-term commitment should you decide you don't like the job, the country, or living abroad. All of the logistical and legal legwork to secure your work and residency permits, housing, job leads, or a job are done by the organization, so you can simply enjoy your time abroad. Some programs will require that you find a job once you are in the country, but they'll give you a work permit and job leads to make the whole job-hunting process easier than it would otherwise be.

RECEIVE COLLEGE CREDIT

Organized programs like approved internships, volunteer positions, or community service work offer college credit toward your degree requirements. Talk to your study-abroad advisor to make sure your overseas work can be integrated into your degree program.

ESTABLISH A FOUNDATION FOR AN INTERNATIONAL CAREER

If you are considering working abroad as part of a larger, global career plan, gaining professional skills and relevant work experience through a work-abroad program can help you evaluate the work environment and other types of career opportunities in that country. If you think you want to be employed abroad for the long term, use your work-abroad program to develop international skills and contacts to leverage in your future job hunt. The self-awareness and maturity you'll gain by living abroad, regardless of the kind of job you have, are positive steps toward self-development and career growth. Work-exchange programs arranged through universities and exchange organizations are designed to offer students solid overseas experience, which will become the foundation for the professional work opportunities discussed in the next chapter.

How to Avoid the Pitfalls of Working Abroad

Working abroad should benefit you both personally and professionally. To make the most out of it, know what pitfalls to avoid. The following guidelines will help you do that.

SELECT A REPUTABLE EXCHANGE ORGANIZATION

Whether you plan your overseas work through your university or an exchange organization, research how long they've been in business, how they arrange work permits (make sure they do it legally), and how much they charge for their services. If the fees seem high or anything about their reputation seems questionable, compare their offers with other exchange organizations. Unfortunately, there are organizations that claim they can get you a work permit and job abroad, but whose business practices are illegal or a scam. The exchange organizations mentioned in this book are some of the biggest and most reputable in the world.

SELECT A WORK-ABROAD PROGRAM THAT MEETS YOUR GOALS

There are lots of different types of jobs you can do overseas. To make the most of your work experience, make sure your work-abroad program closely ties in with your studies, your professional field, or your personal reasons for going abroad. Discuss your work-abroad goals with your study or career advisor. Make sure the program you pick can place you in an appropriate position to gain the exposure and skills you need, whether it be in a hospital, a bank, a government agency, or a nonprofit organization. Also request the accommodations that meet your needs, such as living with a family to get optimal cultural exposure. Your exchange program should also confirm that you can switch your housing or job arrangements if there is a problem.

TALK WITH PROGRAM PARTICIPANTS

Doing a bit of research about your program beforehand can give you some advance notice on what to expect. The best way to get answers to your questions is to go right to the source, that is, the participants themselves. Current and past program participants can give you their impressions of the program, insider tips, cultural advice, and perhaps even the names of friends and contacts in your target country. Feel free to ask candid questions, like if you get to work on relevant work projects and take part in meetings or if you have to make photocopies all day. Also ask participants about their cultural adjustment and whether they thought it was strange to return home after living abroad. You may be surprised at their answers.

Planning Your Work-Abroad Adventure

How do you plan an overseas work program to meet your needs? First decide *how* you want to work abroad and *what* you want to do. To find out about the wide variety of programs available, stop by your career advisor's office or search the umbrella Web sites mentioned below. Work-exchange programs are categorized as:

- Internships and volunteer programs
- Trainee and international management trainee programs
- Teaching positions
- Short-term jobs and au-pair positions

Each of these work-abroad programs will be defined more fully in the next section. Once you find a program that meets your academic, personal, and professional needs, you simply apply. Some programs may have prerequisites and a fee, so be sure to read the fine print.

Getting Started: Using the Internet to Research

There are literally thousands of interesting work, internship, and volunteer programs offered through universities, companies, and work-exchange organizations, so it can be overwhelming to find the best one for you. If you are a student, check with your school's career or study-abroad office to find out what programs they offer and how they can be integrated into your degree requirements. If you're a nonstudent, check with local colleges or universities to see if you're eligible for one of their programs. Some programs accept and place working professionals into international positions. Finally, research the Web sites listed here, as they typically include many small and regional organizations in their search databases.

All Abroad (see page 127)
www.allabroad.com

Alliances Abroad (see page 127)
www.alliancesabroad.com

BUNAC (see page 231)
www.bunac.org

CIEE (see page 127)
www.ciee.org or www.councilexchanges.org/us

GoAbroad/InternAbroad/JobsAbroad (see page 128)
www.goabroad.com or www.internabroad.com or www.jobsabroad.com

Transitions Abroad (see page 130)
www.transitionsabroad.com or www.transitionsabroad.com/resources/work

OTHER RESOURCES

Alternatives to the Peace Corps: A Directory of Third World & U.S. Volunteer Opportunities
9th edition, edited by Joan Powell (Food First Books, 2000)

In this handy little guide, you'll gain insight into the Peace Corps experience and learn how to become an informed and aware volunteer.

The Back Door Guide to Short-Term Job Adventures
by Michael Landes (Ten Speed Press, 2000)

This is a motivating guide that takes a self-enrichment and holistic approach to finding short-term work and volunteer opportunities. In Chapter 11: Abroad Adventures, there are excellent descriptions of a wide variety of short-term jobs, internships, and volunteer programs, and detailed application instructions.

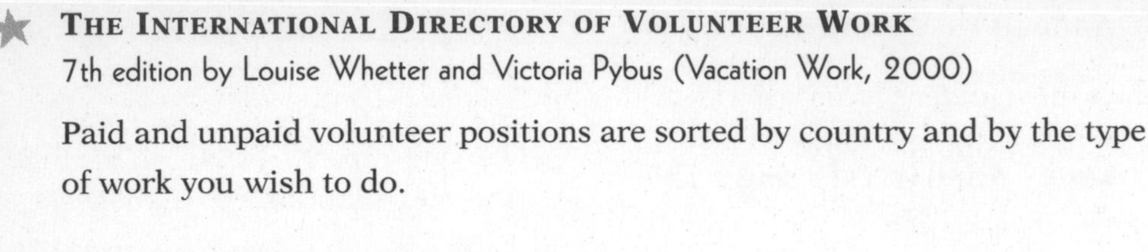

The International Directory of Volunteer Work
7th edition by Louise Whetter and Victoria Pybus (Vacation Work, 2000)

Paid and unpaid volunteer positions are sorted by country and by the type of work you wish to do.

So, You Want to Join the Peace Corps . . . What to Know Before You Go
by Dillon Banerjee (Ten Speed Press, 2000)

If you are thinking about joining the Peace Corps, this book, written by a former Peace Corps volunteer, answers many of your questions and concerns.

Volunteer Vacations: Short-Term Adventures That Will Benefit You and Others
7th edition by Bill McMillon and Edward Asner (Chicago Review Press, 1999)

This book covers about 250 volunteer organizations worldwide that you can search by cost, length, location, season, or program type.

Work Abroad: The Complete Guide to Finding a Job Overseas
edited by Clayton A. Hubbs (Transitions Abroad Publishing, 1999)

This book lists many different programs and resources for internships, volunteering, short-term work abroad, and more.

Vacation Work's Work Your Way Around the World
revised and updated 10th edition by Susan Griffith (Vacation Work, 2001)

This is an indispensable guide for the working traveler that includes information on working in tourism, agriculture, teaching, child care, business, and for volunteer organizations, as well as details about working in European and other countries around the world.

Your Working-Abroad Options

A variety of work-abroad programs are detailed below.

INTERNSHIPS AND VOLUNTEER PROGRAMS

Nowadays most undergraduate and graduate programs encourage students to take part in an internship or volunteer program to obtain practical work experience. This is especially true for international business, international relations, or language majors, and others whose studies are global in nature. Spending a summer working on economic development initiatives in Mexico or a semester studying European politics in Brussels could give you a jump start on a career in global business or politics. Internships and volunteer stints can be coordinated with international companies, governments, universities, schools, nonprofit groups, and many other associations around the world. Your university may have special arrangements with certain work-abroad programs to ensure that your internship or volunteer service complements your degree requirements and supplies credits toward graduation, so start your research at your university's study/work-abroad office. BUNAC and CIEE, listed below, are two of the most reputable exchange organizations in the world and can arrange a wide variety of work-abroad options for students. Other internship and volunteer programs can be found by doing a quick search on any of the Internet Web sites mentioned previously (see page 229). For more information on volunteer programs overseas, check out chapter six.

★ **BUNAC USA**

www.bunac.org

P.O. Box 430
Southbury, CT 06488
Toll-free: 800-462-8622
Phone: 203-264-0901
Fax: 203-264-0251

BUNAC, or the British Universities North America Club, arranges work permits for students and young people to work in England, Scotland, Wales, Northern Ireland, Australia, and New Zealand. To work in the U.K., you must be at least eighteen years old and taking eight credits or more toward a degree. The Blue Card, a special student work permit enabling you to work in paid positions, is valid for six months. The New Zealand program allows people ages eighteen to thirty to work for up to one year. In Australia, people from ages eighteen to thirty can work for up to four months. See BUNAC's Web site for the individual program requirements.

★ **COUNCIL ON INTERNATIONAL EDUCATIONAL EXCHANGE (CIEE)**

www.ciee.org

205 East 42nd Street
New York, NY 10017
Toll-free: 888-COUNCIL

CIEE administers work, intern, volunteer, study, and travel programs for students ages eighteen and up from many countries. Typical jobs range from casual work in the service or tourist industries to professional work in business. CIEE will arrange your work permit and provide program support to find you a job and accommodations.

For other internship and volunteer opportunities, contact Alliances Abroad (see page 127).

TRAINEE AND INTERNATIONAL MANAGEMENT TRAINEE PROGRAMS

Business, economics, or engineering majors who are interested in skilled, short-term employment may be able to arrange a traineeship abroad. Traineeships are designed to build professional skills and are available to students who have completed at least two years of college. The programs vary in length from a few months to one year and usually include compensation sufficient to cover living expenses. If you're no longer a student but are interested in doing a traineeship, check out the Association for International Practical Training's Career Development Exchange Program, which accepts professionals up to the age of thirty-five.

★ AIESEC (International Association of Economic and Management Students)

www.aiesec.org

Phone: 212-757-3774
Call for the location of your local AIESEC office.

AIESEC facilitates international exchanges of students and recent graduates for paid traineeships or volunteers in nonprofit organizations. Exchanges are offered in business management, technology, and development. AIESEC runs the most extensive youth exchange in the world and prides itself on being able to place students in challenging, career-oriented training positions. To apply, call your local AIESEC office for an application and deadline information.

★ Association for International Practical Training (AIPT)

www.aipt.org

10400 Little Patuxent Parkway, Suite 250
Columbia, MD 21044
Phone: 410-997-2200
Fax: 410-992-3924

Through its Career Development Exchange Program, AIPT helps graduates and professionals from eighteen to thirty-five years old arrange professional training overseas. Professionals need to have relevant work experience, students must have relevant course work, and both need to be proficient in the local language. Training programs are offered in the fields of business, culinary arts, architecture, teaching, finance, and more.

The process works like this: You pay AIPT a $75 USD application fee, then you post your résumé on their online placement service and apply for jobs in their database of employers. When a match is found and you receive an offer, you have seven business days to decide whether or not to accept it. Once you accept an offer, AIPT arranges the work permit. Once you have your permit in hand, you are on your way to work overseas! AIPT can currently arrange a work permit for you in Germany, Switzerland, the United Kingdom, France, Brazil, Mexico, Finland, Ireland, and Sweden. Training positions receive at least thirty-five hours of paid work per week and will not exceed eighteen months.

Also, through the International Association for the Exchange of Students for Technical Experience (IAESTE), AIPT offers placement assistance to students with technical majors like engineering, computer science, and chemistry.

★ The Carl Duisberg Society/Carl Duisberg Gesellschaft (CDS) Internship and Career Training Programs

www.cdsintl.org or www.cdg.de

871 United Nations Plaza, 15th Floor
(First Avenue at 49th Street)
New York, NY 10017
Phone: 212-497-3500
Fax: 212-497-3535

CDS offers college students, recent graduates, and young professionals under thirty several ways to gain professional work experience. CDS issues work permits valid for up to eighteen months and will assist you with finding a job. The Internship Program and Summer Internship Program are for students or grads pursuing their first international work experience. The Career Training Program is for students or grads who already have some previous work experience. CDS issues a work permit for six months (or three months for the summer) and arranges a paid internship for you in your field. Summer internships are available in Germany, Turkey, and Argentina. Independent work-abroad programs are offered in Germany, Switzerland, and Singapore. CDS also arranges scholarship/fellowship programs, such as the Robert Bosch Foundation Fellowship, which offers American professionals up to the age of thirty-four the opportunity to partake in a nine-month executive-level internship in the federal government and private sector.

Interning in Germany

by Elizabeth Kruempelmann (ekruempe@hotmail.com)

After my first study-abroad program in Denmark and an intensive language course in Germany, I did all kinds of research on ways to get myself back to Germany to learn the language fluently and gain international work experience. What I found was the Carl Duisberg Society (www.cdg.de).

The career training internship I did through the Carl Duisberg Society's (CDS) exchange program was a turning point in my personal and professional growth and my general life direction. CDS provided me with the proper work and residency permits, a one-month language course, and a family stay. Finding a job or internship and a permanent place to live were my responsibilities.

It was far from easy to get a job with hardly any work experience, rudimentary language ability, and the worst unemployment rate in Germany since World War II! I hadn't expected all these little difficulties: trouble finding a job, problems communicating in German, issues with the German mentality, sickness from the cold and dreary weather. My glorious illusions of life in Europe dissolved within the first few months—and I wallowed in a pit of depression, waiting in pathetic despair for something good to happen. Despite my previous study-abroad and travel experience—where the logistics were mostly preplanned—my experience in Germany was hard at first. I hated it, I cried, but I never gave up (although I thought about it). I stuck it out every day.

Then finally, lightning struck, the telephone rang, and KPMG Management Consultants offered me a position as an intern. It was an interesting opportunity to get business experience, practice my business German, and learn about the consulting business. Once I had a job, rather miraculously, life started to get better. I spoke German with more ease, the mentality of the German people became a new and enlightening perspective on the world, spring came, people smiled, and I suddenly realized I had a whole group of fun, international friends, including a German boyfriend. I made it through the hardest part—adapting to German ways—and started to enjoy the language, culture, and social life. When my internship ended, I decided to extend my work and residency permits for another nine months (during which I taught English).

As I reflect back on my experiences studying and working in five countries, learning three foreign languages, and traveling to thirty lands, I realize that the CDS program was my basic training for an international life. My experience in Germany, although difficult at first, taught me what I needed to know about myself to be successful in the world. I was constantly questioning my goals and the means I'd chosen to reach them. Do I stay or do I go? Do I quit a job I hate or do I do what's necessary to get to the next step? I had to peel off several layers of stereotype and prejudice to make room for all the new knowledge I acquired. And though the steep learning curve I endured at the beginning often felt like "boot camp," I did eventually learn how to survive on my own. And was it ever worth it!

Read more about Elizabeth Kruempelmann's international experiences elsewhere in this chapter as well as in chapters two, five, and eight.

If you have the right credentials, an international management-trainee program is another option for working overseas. These programs are typically offered directly by companies, or as part of a college recruitment process, and rarely through a work-abroad program. Multinational companies are looking for graduates and MBAs who have the potential to take on leadership positions. Some of these companies have their own management-training programs, which often include international rotations. Most are very competitive, but if you are destined for a corporate career path, being accepted into an international management trainee program could set you up for a successful global career in the corporate business world.

Most international banks, consulting companies, and multinational corporations have career-starting opportunities like internships and management-trainee programs. Research the Web sites of companies that interest you for information on management-trainee programs with an international focus. Not all companies send new hires abroad to train, but some training programs may involve global rotations. If you already live overseas and are looking for a job, try to arrange a management trainee program in that country. Also, ask career counselors at your university or MBA program which companies actively recruit and offer international training.

The following organizations actively recruit management trainees.

★ Citibank/Citigroup

www.citigroup.com/newgrads/

Citigroup offers undergraduates and graduates a variety of training programs, some of which are global in nature or which offer global rotations.

★ GE Capital

www.gecaptial.com

The International Global Leadership Development Program (GLDP) from GE Capital is an elite program where the "best of the best" are chosen from limited U.S. universities and international business schools. GLDP provides its participants with strong technical and strategic skills for global careers at GE Capital.

★ International Training Centre for Women (ITW)

www.euronet.nl/users/itw/

ITW is designed to address the needs of women worldwide in the areas of management, entrepreneurship, training, skill development, consultancy, and research. They offer a four-week core program and a six-week extended course in the Netherlands taught in English.

TEACHING ABROAD

There are many opportunities for teaching overseas. Depending on your subject of interest, you can develop a career teaching English on a freelance or permanent basis, or you can teach at international schools or local universities. If you already have a few years of teaching experience, you might also consider a teaching exchange. Requirements for teaching abroad vary depending on the position. In some cases, a bachelor's degree, a teaching certificate from a recognized U.S. college or university, and one year of teaching experience are required. In other cases, particularly TESL (Teaching English as a Second Language) or TEFL (Teaching English as a Foreign Language) instruction, you may only be able to teach with your TESL

A GLOBAL CITIZEN'S PERSPECTIVE

Citibank's Management Trainee Program

by Sarah Seeland (sseeland@yahoo.com)

I made the commitment, after leaving the United States on a one-way ticket, to work in Germany—no matter what. I would have become an expert on fine German beers by working as a bartender or waitress, but what I really wanted was experience I could put on my résumé to build my credibility. I discovered that participating in a management training program was an excellent way to do that.

Many large international corporations recruit new graduates for training programs with the intention of grooming students to become future employees. I was offered a six-month management associate trainee program with Citibank in Frankfurt, which included extensive training in London focused on finance, banking products, and management. I rotated through all of the functional departments within Citibank's corporate banking operations. After I completed my six months of training, I was recruited within the bank to manage the Securities Clearing and Global Custody teams. Included in my benefits package were six weeks of paid vacation time, relocation expenses, apartment-finding assistance, and my salary. I was given the opportunity to manage a team of six, which provided eight hours of intensive German language and work environment study a day. While some expatriates land positions that allow them to speak English, the fact that I was only able to communicate in German accelerated my knowledge of the language and culture. After a rewarding three-year career with Citicorp in Frankfurt, and three and a half years in Germany, I returned to the United States to continue my journey.

I ended up moving to San Francisco and made a job leap from banking to technology. While working for the Oracle Corporation, I was given the opportunity to move to New York to join the Citigroup sales team because of my experience working with Citibank in Frankfurt. From there I moved to EMC Corporation and continued to sell technology to Citigroup. My international experience with Citicorp was a key ingredient that allowed me to change careers from banking to technology, and change positions from product management to sales. The professional credibility and personal self-confidence that my experience abroad gave me is priceless.

or TEFL certification. In addition, an ability to speak the country's native language may not always be required.

Teaching English

One of the easiest and most popular ways to work abroad is to teach English. In most countries where English is not the mother tongue, the demand for English instruction at all levels is great. Your opportunities will be enhanced if you have a degree or certificate in Teaching English as a Second or Foreign Language (TESL or TEFL) and previous teaching experience.

★ Dave's ESL Café

www.eslcafe.com

"The Internet's meeting place for ESL/EFL students and teachers from around the world" has a host of information from chat sites to teaching tips.

★ ESL Magazine Online

www.eslmag.com

Aimed at ESL professionals, *ESL Magazine* is an e-zine with articles and resources for teaching English overseas.

★ International House CELTA Program

www.ih-portland.com

320 Wilshire Blvd., Third Floor
Santa Monica, CA 90401
Phone: 310-394-8618
Fax: 310-394-2708

International House (IH), the world's largest teacher-training organization, has offices worldwide. CELTA (Certificate in English Language Teaching to Adults) is one of the most widely recognized entry-level teaching qualifications in the world. IH offers a four-week training course, practical training, and a job-placement service.

A Global Citizen's Perspective

Teaching English in Germany

by Elizabeth Kruempelmann (ekruempe@hotmail.com)

When I was pursuing ways to work abroad, teaching English never even entered my mind. I didn't think it was the type of professional experience I wanted at the time. But I couldn't have been more wrong. It wasn't until my internship in Germany came to an end, and I was nearly desperate to find a way to stay in Germany, that I gave it a passing thought.

Even though I had a work permit, none of the language schools for whom I worked ever asked to see it. Because the language schools needed native speakers, and local workers did not qualify for these positions, the work permit wasn't an issue. After filling out a job application at Berlitz Language School, I was contacted the next day and told that if I wanted to teach I would have to participate in a two-week course to get certified in the Berlitz teaching method. Other than a bachelor's degree, I didn't need any other teaching qualifications or ESL certifications. After I completed my Berlitz course, I was qualified to teach at any Berlitz school in Germany or around the world.

As a freelance teacher and therefore not a full-time employee of Berlitz, I had the freedom to arrange my schedule as I wished. Although the salary was decent, 25 DM per hour (or approximately $18 USD at that time), the hours were unpredictable. So, I decided to get a second teaching job at another local language school. I even started a side business giving private lessons to students and professionals.

The language schools contacted me every week to arrange my work schedule. Sometimes I would have to pull myself out of bed after a night out at the local *Kneipe* (bar) to make it to an 8:00 A.M. class. Other times I would spend my days biking along the Rhine River, getting back to town just in time to teach the after-work crowd.

I taught conversational classes and business English in a variety of international companies and in the classroom. Classes ranged from one student in a private, intensive all-day class five times a week to groups of six to fifteen people in a regular hour-and-a-half class twice a week. The people I met were fabulous and the satisfaction I felt was rewarding. It made my day to see my students' eyes light up in excitement when they were able to say something in English. The biggest challenge was teaching savvy multilingual professionals, like the Swedish secretaries from Ericsson, whose English was nearly flawless!

Surprisingly enough, my students were not only Germans, but an eclectic mix of people from all over the world: a young expatriate couple who worked at the Japanese embassy, German factory workers, and university students of various nationalities. As it turned out, teaching English was one of the most fulfilling professional experiences I could have hoped for at the time. It allowed me to earn more money than I did in my internship, and the flexible work schedule gave me the freedom to take some cool trips to Amsterdam, Paris, and other towns in Germany.

Read more about Elizabeth Kruempelmann's international experiences elsewhere in this chapter as well as in chapters two, five, and eight.

★ I-TO-I

www.onlinetefl.com

One Cottage Road
Headingly
Leeds LS6 4DD
United Kingdom
Phone: 44 (0) 870-333-2332
Fax: 44 (0) 113-274-6923

As one of the United Kingdom's leading TEFL (Teaching English as a Foreign Language) organizations, i-to-i offers an online TEFL course consisting of ten Web-based modules, which is similar to a four-week TEFL course. You will receive a CD-ROM, a hard copy guidebook of the course, and a TEFL certificate upon completion. The cost is approximately $295 USD, but you need to pay in British pounds according to the current exchange rate. i-to-i also coordinates volunteer teaching, conservation, and business placements in twelve countries.

★ JAPAN EXCHANGE AND TEACHING PROGRAM (JET)

www.mofa.go.ip/l_info/visit/jet/index.html or for U.S. citizens: www.embjapan.org/jet

50 Fremont Street, Suite 2200
San Francisco, CA 94105
Toll Free: 800-463-6538
Phone: 415-356-2462
Check the Web site for a JET chapter near you.

The JET program increases foreign-language education in Japan and promotes international contacts at the local level by fostering ties between Japanese youth and young foreign graduates. JET gives graduates the opportunity to teach English in Japan or participate in international exchanges for language education.

New World Teachers

www.goteach.com

605 Market Street, Suite 750
San Francisco, CA 94105
Phone: 415-546-5200
Fax: 415-546-4196

New World Teachers is America's largest TEFL/TESL (Teaching English as a Foreign/Second Language) training program. It trains and offers its teachers assistance in finding jobs overseas teaching English. Extensive Internet resources and a two-day seminar in Teaching English to Young Learners are available with every course.

Teachabroad.com

www.teachabroad.com

Teachabroad.com allows you to search through a directory of educational opportunities and paid and volunteer teaching positions overseas.

TESOL Placement Services

www.tesol.edu

700 South Washington Street, Suite 200
Alexandria, VA 22314
Phone: 703-836-0774
Fax: 703-836-6447

TESOL Career Service is the best place to find TESOL (Teaching of English to Speakers of Other Languages) teaching positions worldwide. The TESOL *Placement Bulletin* (ten issues) has worldwide job listings and job-searching resources. The Employment Clearinghouse at TESOL's annual convention in March is the world's largest ESOL job fair, where more than one hundred schools recruit applicants on site.

★ Transitions Abroad

www.transitionsabroad.com

The Transitions Abroad Web site offers a listing and description of ESL (English as a Second Language) training and placement programs.

★ Teaching English Abroad: Talk Your Way Around the World!

5th edition, by Susan Griffith and Victoria Pybus (Vacation Work 2001)

This book provides up-to-date contacts, specific job vacancy information, and real-life examples of people who have taught overseas.

★ Teaching English Overseas–A Job Guide for Americans and Canadians

by Jeff Mohamed (English International, 2000)

This is a guide for Americans and Canadians that includes advice on teaching without training, choosing a training program, conducting a successful job search, and dealing with culture shock.

For more information about teaching abroad, check out the Peace Corps (see page 209) and WorldTeach (see page 212).

Teach at American or International Schools Overseas

Another teaching option is to work at an American or international school overseas. International schools need a wide variety of teachers, including English teachers. Qualified K–12 teachers can find out about positions in U.S.-style international and elementary schools abroad by attending a recruiting fair. International School Services (ISS) organizes recruiting fairs in various cities throughout the year.

★ Department of Defense Education Activity

www.state.gov/www/about_state/schools/index.html

4040 N. Fairfax Drive
Arlington, VA 22201

The Office of Overseas Schools/U.S. State Department is a key site to refer to if you want to teach abroad. You can find information on American-sponsored elementary and secondary schools overseas, a job-hunting link, recruiting agencies, and organizations with teaching opportunities.

The Department of Defense Dependents Schools (DoDDS) is a worldwide school system from prekindergarten to grade 12 operated by the U.S. Department of Defense in fourteen countries. The Department of Defense employs 6,500 educators to teach 77,000 students in 157 elementary, middle, and secondary schools. A wide variety of positions are available, from teachers to school nurses and psychologists. For teacher application requirements and procedures, refer to the Department of Defense Education Activity/Employment opportunities at the State Department's Web site.

★ International Schools Services (ISS) Recruiting Fair

Educational Staffing

www.iss.edu

P.O. Box 5910
Princeton, NJ 08543
Phone: 609-452-0990
Fax: 609-452-2690

Recruiting fairs take place in the United States for K–12 teachers who are seeking positions with elementary and secondary schools that employ American educators abroad. Applicants must have a bachelor's degree and two years of current K–12 teaching experience or overseas living experience. Contact ISS for specific fair dates and locations. Credential files must be received six weeks before fair dates.

Participate in a Teacher Exchange

A final option for teaching overseas involves formal exchanges through which U.S. and foreign teachers switch positions, and retain salary and benefits from their home schools. The Fulbright Teacher and Administrator Exchange Program (see page 156) is one of the most popular.

SHORT-TERM WORK AND AU-PAIR PROGRAMS

Working for the summer or for a short period of time is an excellent way to get your feet wet without making a long-term international commitment. Summer and short-term work mostly consists of jobs in the service, tourism, or agricultural industries, such as bartending or working on a farm or kibbutz.

★ International Cooperative Education

www.icemenlo.com

15 Spiros Way
Menlo Park, CA 94025
Phone: 650-323-4944
Fax: 650-323-1104

International Cooperative Education arranges two- to three-month paid work and internships in retail sales, hotels, restaurants, au-pair situations, agriculture, education, banking, and business. Students must be under the age of thirty and some programs require a language background. Check the Web site's database of job listings.

Au-pair work is essentially a nanny job that lasts a few months to a year or longer. Generally you will live with a family, watch their children, and take part in their daily life. Hours can range from a few up to twelve per day, and usually you'll be given one or two days free. Most of the time you receive a modest allowance that you can use as spending money. Room and board are paid by the host family, but the au pair is responsible for transportation to the country. If you like children and get along easily with people, au-pairing has many rewards and offers a way to experience a culture from an inside perspective. Au-pairing is a particularly good option for female high-school students, but many positions are available for older women as well.

★ Au Pair in Europe

www.princeent.com/aupair

P.O. Box 68056
Blakely Postal Outlet
Hamilton, Ontario L8M 3M7
Canada
Phone: 905-545-6305
Fax: 905-544-4121

Despite the name, this agency places au pairs aged eighteen to thirty in twenty-one countries, including Australia, Iceland, Russia, Japan, New Zealand, and Monaco. Au pairs generally work about thirty hours per week, and receive a weekly stipend. Contracts last from three to twelve months in duration. Au pairs pay a placement fee and transportation to the country, then the family will pay for most expenses.

★ AuPair Wizard

www.au-pair.com

Europublic Werbeagentur GmbH
Neue Schoenhauser Strasse 10
D-10178 Berlin
Germany

AuPair Wizard is a service from au-pair.com that allows au-pair applicants from various countries to apply for au-pair positions free of charge. AuPair Wizard has listings of au-pair institutes in sixteen countries in Europe, Canada, the United States, Africa, and the Middle East. To apply, simply enter information about your background, including your top three preferred countries, and then submit the data to the requested agencies for processing. Applicants will be contacted with details of potential au-pair opportunities. Families looking for au pairs may also use this service.

CHAPTER EIGHT

Working Abroad Alternatives: Building Professional Skills and a Global Career

The people who get on in this world are the people who get up and look for the circumstances they want, and, if they can't find them, make them.

GEORGE BERNARD SHAW

People from a variety of backgrounds find overseas jobs in good and bad economic times through persistence, connections, proper preparation, and luck. It is not always easy to find work opportunities abroad; however, instead of leaving your future up to fate, you can take steps to significantly increase your chances of landing the right position in your dream destination.

Hunting for an international position involves different strategies than searching for a position in your home country does. As you start to research your options, you'll be confronted by the "International Job-Market Realities" discussed in chapter seven (see page 218). In many (but not all) cases, getting hired by an international company requires that you have international experience in your field. That's why internships, volunteer service, and trainee programs can be essential first steps to breaking into the international job market. And once an international firm hires you, you may have to spend a few years working your way into the international department or a position that will send you overseas.

If you manage to get hired by a company that will send you abroad, the company will usually take care of arranging for your work permit. However, if you go overseas on your own to get a job or decide to work for yourself while you are abroad, you will be faced with solving the work-permit problem on your own. Securing a work permit or finding a way around the work-permit problem (see page 220) will put you on your way to gaining the overseas work experience that will be the foundation for a successful global career.

The Benefits of Working Abroad

This chapter will tell you how to start your search for international work experience and describe the many benefits you will reap from working abroad.

OPPORTUNITY COMES KNOCKING

One of the benefits of working abroad is that many opportunities will come your way. The trick is learning to recognize an opportunity when it presents itself. Becoming an opportunist—having the ability to turn potentially negative situations into positive experiences—is one of the most beneficial life skills you can learn by working abroad. Securing a job overseas and working in a foreign environment won't be as easy as it is at home, but learning to make the most of every opportunity is where challenge, fun, and wisdom come into play.

To actively seek a job overseas, you will have to educate yourself about the necessary work permits, residency permits, working conditions, language requirements, salary levels, and a host of other bureaucratic issues. You may have to sell yourself aggressively to companies who might be tempted to hire a local who speaks four languages, has an international MBA, and several years of international work experience. But don't let these issues intimidate you. Getting yourself hired in the international environment, like starting any new job, involves a steep learning curve at the beginning and is simply the reality of working abroad. But once you learn to navigate the global workplace, you'll never forget how to do it. Every challenge you encounter uncovers a potential new life path. Learning to take advantage of these opportunities will give you the survival skills you need to succeed anywhere—at home and abroad.

GAIN WORK EXPERIENCE AND PROFESSIONAL SKILLS

Students, professionals, and career changers can gain exciting international work experience and skills through internships, short-term work exchanges, volunteering, or working full-time for a company overseas. If you are really ambitious you could succeed in building a career consisting of an internationally recognized degree, fluency in multiple languages, work experience in several different countries, international contacts, and job offers worldwide. Studying abroad has become a global trend and as the Internet internationalizes the job search process, more and more people will find a world of work opening to them.

ACQUIRE SURVIVAL SKILLS

How do you survive while looking for a job abroad? You must stay confident, persistent, and flexible. These are the essential skills you'll need to reach your goals at home or abroad. Confidence means keeping your self-esteem high at all times, especially during the rough patches. Lift your spirits by exercising, thinking positively, and meeting people in your new environment. Persistence means staying focused, doing something every day to put you one step closer to your goal. Be patient enough to do what it takes to reach your objectives; take time to educate yourself about how business is done abroad so you can minimize your cultural blunders. Flexibility means doing something that you might not ordinarily do. Live in a new country, learn the language, make contacts, and continue to pursue your ambitions. Have realistic expectations of what it takes to reach your goal (this book will help you learn what those are). When faced with a "sink or swim" situation, your priorities might change or clarify, resulting in a fantastic experience you hadn't dreamed of.

UNDERSTAND CULTURAL AND BUSINESS DIFFERENCES

Working abroad teaches you to recognize and appreciate cultural differences and understand alternative ways of doing business. As your contact with the international community increases, you may find that you have things in common with people you wouldn't ordinarily seek out. It's rewarding to watch your cultural and business training pay off when you know how to bow properly to your new Japanese boss, negotiate a deal using business German and proper etiquette, or greet your Portuguese female friends with a kiss on each cheek. You never know when these skills will be needed in the future.

NEGOTIATE SUCCESSFULLY WITH A FOREIGN CULTURE

Negotiation skills are a highly valued professional skill in any business, and having a good command of the local language only improves your career chances. Although English might be used quite extensively in international business, being able to negotiate in the local language will earn others' respect and provide you with insight about doing business with that culture. If a Spanish businessperson tells you in English that he'll give you a report "tomorrow," it likely means something different than if a German colleague told you the same. It is critical to understand the cultural interpretation behind the words since they could affect your negotiation tactics. If you work abroad, some negotiations will take place in English, especially if other foreigners are present. However, your sensitivity to cultural issues will often tell you more about a client than their words do.

BUILD AN INTERNATIONAL NETWORK

Networking and then utilizing your resources is essential to creating job opportunities. Your international travels will expose you to a variety of social and business situations—perfect opportunities to build a new circle of multicultural friends and business contacts. You never know when they might be able to assist you or you might have the resources to help them.

BECOME INTERNATIONALLY SAVVY

Even if you already have professional work experience in the States, there are still important things to learn about doing business overseas. You may be business savvy, but are you internationally savvy? Working abroad gives you a chance to gain the international skills you might heretofore have been lacking and puts you on par with your European, Asian, African, and South American counterparts.

How to Avoid the Pitfalls of Working Abroad

There are many benefits of gaining professional work experience overseas. Some people will receive a good job offer rather quickly, and others may never get the offer they were looking for. It all comes down to having realistic expectations, maintaining a positive attitude, and learning to make the most out of your experiences.

KEEP YOUR EXPECTATIONS REALISTIC

One of the most common first mistakes I see job hunters make is to expect a full-time job overseas that is comparable in salary, benefits, and job satisfaction to similar jobs at home. Although it is possible to be sent abroad with a fabulous relocation package full of perks (and you'll know if you qualify for this), the number of international job seekers out there far outweighs the number of cushy expat opportunities overseas. Having said that, if you do get an offer (or even multiple offers) to work overseas, you might encounter salaries that are substantially higher or lower than what you are used to. How do you evaluate foreign salaries, cost of living, and the whole expat deal? Check out the section in this chapter titled "Find a Job with a Company That Will Send You Overseas"(see page 257) for advice.

The point is to keep your expectations realistic and know what you are willing to sacrifice to work abroad. Many job hunters claim they are only looking for professional full-time employment with benefits (like they can secure in their home country) until they are hit with the hard reality of high unemployment, difficulty obtaining work permits, and fierce competition. Those who get discouraged are the ones who do not know the realities of the global marketplace, do not have the skills that will set them apart from the competition, or don't know how to market themselves in a foreign environment. Although this sobering reality often comes as a tremendous shock, the jolt will inevitably force determined international job seekers to reexamine their reasons for going abroad and solidify their resolve.

SORT OUT YOUR LIFE AND CAREER PRIORITIES

Figure out the main reason you would like to go overseas. What stage of life are you in now? Do you want to continue on your current career path within a business, government, or nongovernmental organization; get expatriate benefits (see page 270) and pursue your career in several different countries; obtain international work experience and see the world; learn a foreign language; or be with a significant other and support yourself financially? Maybe you have a reason for going abroad that I haven't mentioned here. Know your priorities from the start and figure out how flexible you are willing to be. (See chapter three for more tips on setting your goals.)

Planning Your Work-Abroad Adventure

To build a global career you may have to get an advanced degree, enhance your professional experience, or become proficient in a foreign language. Consider the work a long-term investment in your future. Here are some of the ways you can prepare for a global career.

INTERNATIONAL BUSINESS

If you want to work for a multinational company, a bank, or a consulting firm in an international capacity, a graduate degree—preferably an MBA (see the MBA school listings in chapter five)—and international work or internship experience abroad will put you in the most competitive position. Whatever degree you decide to get, it is imperative that you can prove that you have strong finance and marketing skills, some technical knowledge, and foreign language fluency. Your skill set must be on par or better than your foreign counterparts, who have MBAs from international schools, business and technical knowledge, fluency in multiple languages, a worldwide perspective, and who may ultimately cost companies less to employ than an American. To help young professionals get started, many international companies, banks, and consulting firms offer internships, graduate training programs, and international career paths. Check the Web sites of the companies that interest you, talk with people who work there about international opportunities, and contact your university's career office to see which global companies recruit on campus. As international business spans many fields, also consider global jobs in advertising, with Internet companies, in journalism or broadcasting, or in other areas where the hiring requirements may not be as stringent as those with multinational, Fortune 500 companies. See the "Industry Links to International Jobs" section on page 291 for more information.

GOVERNMENT JOBS

The most common jobs available overseas are with the Foreign Service, the Peace Corps, the U.S. military, and the CIA. However, there are many government departments and agencies with global-oriented positions stateside that enable you to work on foreign policy, with foreign trade and commerce, or in intelligence. These types of positions, although global in scope, may never offer you the chance to work in a foreign country on a long-term basis. Don't depend on a government job to give you work experience abroad. (In fact, you are often expected to bring to the position several years of overseas experience, language fluency, a graduate degree, and professional skills.) Luckily, many government agencies also offer internships and programs for young professionals that provide some initial field experience. Although some government agencies have their own Web sites and job postings, many jobs are routed back to the main job bank at the U.S. Office of Personnel Management. See the "Industry Links to International Jobs" section on page 291 for more information.

INTERNATIONAL LAW

Lawyers with a business, finance, or international affairs background and solid language skills are in demand worldwide. High caliber and internationally savvy lawyers may have a choice of job opportunities; however, a master's degree, in addition to a J.D., is helpful for building a legal career with a bank, a law firm, a multinational company, or even a small nonprofit organization. The American Bar Association is a good place to start your research. Also check out the "Industry Links to International Jobs" section on page 291 for more information.

INTERNATIONAL ORGANIZATIONS

International organizations, like the United Nations and World Bank, are very competitive to get into, but worth a try if you have the necessary background. A master's degree or Ph.D. and fluency in multiple languages are usually the minimum requirements for most professional positions. Recent graduates can apply for volunteer positions, internships, or junior posts. See the "Industry Links to International Jobs" section on page 291 for more information.

HEALTH CARE

International health-care issues can range from controlling the spread of AIDS and other life-threatening diseases to women's issues and population control. Skilled doctors, nurses, dentists, and other health-care experts are needed throughout the world. See the "Industry Links to International Jobs" section on page 291 for more information.

NONPROFIT ORGANIZATIONS

Nonprofit organizations work internationally in many ways, including global education, the environment, human rights, and democracy. International positions are available with associations, foundations, government agencies, social service groups, and think tanks (research organizations). As with business and government positions, you will need the right combination of education and professional and international experience to be sent overseas. Contact nonprofit organizations directly to find the exact requirements for overseas positions. See the "Industry Links to International Jobs" section on page 291 for more information.

FRANCHISING OPPORTUNITIES

If you have business experience, an entrepreneurial spirit, and some cash to invest, international franchising opportunities exist in many exciting sectors: restaurant chains such as TGI Friday's, beauty and gift shops such as Crabtree & Evelyn, and language schools like Berlitz. See the "Industry Links to International Jobs" section on page 291 for more information.

Your Alternative Working-Abroad Options

To build a long-term international career either in a foreign country or in your homeland, you'll approach your job hunt in one of four different ways. Find a position with a company or organization that will send you to work in a foreign country, go overseas on your own to get a job, create your own overseas career opportunity instead of placing yourself at the mercy of the local job market, or build your global career with internationally oriented organizations in your home country. This chapter takes an in-depth look at all of these options and gives you the resources you need to make your international career a reality.

FIND A JOB WITH A COMPANY THAT WILL SEND YOU OVERSEAS

One of the best ways to move overseas is to get transferred by your company. They handle your work permit, pay for your moving expenses, and sometimes even find you a place to live. Sounds ideal, right? That's why these positions can be so difficult to find. But it *is* possible. Your first option is to get hired by a multinational company at home that will eventually transfer you to one of their foreign offices. The second option is to get hired by a foreign company that will do the same. Many of the international job seekers I advised while working as a career mentor at Monster.com were hoping to find one of these options.

Being sent abroad by an employer is widely perceived to be the easier, more profitable, and more desired way to find work in a foreign country for the reasons I mentioned above. If you have the right combination of academic credentials, professional experience, and luck, it *can* happen. However, be aware that it is very expensive for companies to send employees overseas and most only send dedicated employees who have shown their commitment to the company and who have proven professional success.

Your best chance of getting sent abroad is to work for an international consulting company, an international bank, a multinational company, or in the hotel industry. You can also find international government positions with the Foreign Service, the CIA, the Defense Department, and with the organizations listed on page 255.

For more information about international jobs, pick up a copy of the following books.

★ **THE COMPLETE GUIDE TO INTERNATIONAL JOBS & CAREERS: YOUR PASSPORT TO A WORLD OF EXCITING AND EXOTIC EMPLOYMENT**
2nd edition, by Ronald L. Krannich, Ph.D. and Caryl R. Krannich, Ph.D. (Impact Publications, 1992)

The authors give a good overview of the international job market, job-search strategies, and types of international work. For more of the Kranniches' books on overseas work opportunities, see www.impactpublications.com.

★ **INTERNATIONAL JOBS: WHERE THEY ARE, HOW TO GET THEM**
5th edition, by Eric Kocher and Nina Segal (Perseus Books, 1999)

This book offers a very good description of international opportunities in government, international organizations, business and banking, nonprofits, international communication, teaching, and law. I especially recommend it if you are trying to choose an international career direction.

To be sent abroad by an employer, you generally have to strategically position yourself, then wait until the right opportunity arises. Being sent abroad by a company is more likely if you have a graduate degree, an MBA, or previous international experience. Although landing a job that will send you abroad with an expat deal can be hard to come by, these tips will get you started.

- Start Where You Are Now

  Ask your current employer what you need to do to be sent overseas with your company. Get involved with the international side of your business. Try to establish contact with foreign colleagues in offices overseas. Utilize your network (see more about networking below). You might even consider a professional exchange program (see page 166).

  If you are a student or recent graduate, start with your university's study-abroad office to arrange an internship or job through a work-exchange program (as discussed in chapter seven). Undergraduate and graduate students can research international companies that recruit on campus, or look into international management trainee programs. Career changers should start with a career assessment. Perhaps your best track involves going back to school for an international degree, studying your specialty overseas, and eventually interning abroad. Alternatively, you can join the Peace Corps, teach, or volunteer overseas, gaining international experience that will put you on the road to a long-term global career. You may also be able to apply the skills you acquired in your last job to a new career in a different industry. Try to obtain international experience in your home market first. That experience could then catapult you into a career with an overseas company.

- Prepare Your Résumé or Curriculum Vitae

  Regardless of where you apply, you will need to prepare a résumé or curriculum vitae (CV), along with a cover letter. To present your education, skills, and strengths in the best possible light, you should be aware of what different cultures expect from job applicants. Americans generally want your résumé to be short, simple, and to the point. Germans prefer their version of a résumé, called a *Lebenslauf,* to be lengthy and detailed, including lists of any diplomas, awards, scholarships, or documents you've ever received to add credibility to your portfolio. If you can, engage the help of a local to help you put your résumé together properly. And if you write it in a foreign language, make sure you have a native speaker proofread it. There are many books on the market that will guide you

Working as an Expat

by Bridget Johns (Bridget_johns@hotmail.com)

After spending fifteen months volunteering with the MBA Enterprise Corps in Poland (see page 187), I started working for a major home-textile manufacturing company in New York City. I eventually ended up at my current company through a contact I made in Poland. It wasn't my ideal job, but I knew it would lead to other opportunities because the international division of my company was quickly growing.

Although I was still based in the United States, my job was internationally oriented. I worked very closely with our European office, as well as offices in Mexico, Canada, and South America. My current position is based in London. My work permit was handled through an agency contracted through my company, and I negotiated my own expat package.

My experience in Poland prepared me for the diverse economic conditions that exist in the regions in which I've worked, including cultural differences and general expectations that non-U.S. customers have of U.S. managers and suppliers. For example, in Europe my colleagues are sometimes taken aback by my direct, "let's get down to business" attitude; however, because this attitude is expected of me, I tend to maintain it. In South America, however, I try to adopt aspects of the local culture, like having three-hour lunches with prospective clients or business partners in an attempt to build solid relationships. For a lot of my international contacts, particularly in Mexico and South America, relationships are often as important, if not more important, than the bottom line.

One of the best things about having an international career is how much of the world you get to see. I've worked with really amazing people and I've seen some of the most amazing sights in the world, but I've also seen poverty beyond imagination. Nothing can prepare you for the abject poverty you encounter in Mexico: small houses built of cardboard or tin located within minutes of wealthy neighborhoods or spectacular natural beauty. When I was in Brazil for a business trip I visited a friend in Rio. We were having dinner along an amazingly beautiful stretch of beach overlooked by hills full of thousands of sparkling lights. When I asked my friend what the beautiful lights were, he explained that the hills were full of *favellas*—shacks made of found materials that housed entire families. The favellas had limited electricity, lacked plumbing, and were completely at the mercy of Mother Nature.

Living and working abroad has taught me to look beyond the surface beauty of a culture to get at the reality beneath. Discovering the true beauty of a land is what keeps me focused and continually seeking my next overseas assignment.

If you know where you want to live, but don't know exactly what you want to do, I suggest moving to the country to look for a job. In my own experience, and through talking to others, I've found that it is very hard to find a job in a foreign country unless you live there.

On the other hand, if you know exactly what you want to do and aren't particularly choosy about what country you live in, find a company in your home country that will send you abroad. It is much easier to get a work permit before you move to a country than after you are already there and working. Also, if you are patient, sometimes the job you've been looking for presents itself within your current organization. As hard as it is to believe, most people aren't very excited about packing up their lives and moving overseas.

If you like the company you work for and there is international activity, be patient. Building international businesses typically takes much longer overseas than it does in the United States. There are geographic boundaries, language issues (even between the United States and the United Kingdom), and cultural differences to take into account before you even start talking about the work. If you are interested in working internationally, but don't work in an international division, tell your boss you'd like to get some exposure to the global side of your business. Perhaps you can work on a global project or volunteer to sit on a committee. Make sure you know the people responsible for your company's international strategy and development and build relationships with them. With patience, a good attitude, and the willingness to listen and learn, an international career can bring rewards beyond your expectations.

Read more about Bridget Johns's international experiences in chapter six.

A Global Citizen's Perspective

Networking for an International Job

by Laurel Eismann (leismann@hotmail.com)

Growing up in a small town in the middle of Kansas, I knew I wanted to travel from a very young age. Luckily for me I chose a profession (computer analyst) with skills that are in demand all over the world. There are many ways to reach your goal of living and working abroad, but I think the most important thing I learned is how to network.

A fellow university student and former work colleague helped me find my first position with an international company that sent me overseas. Learning of my wish to work abroad, this friend referred me to the headhunter who had helped him obtain his position with a company called American Management Systems (AMS) in Denver.

The headhunter had worked with several international companies, but figured I would also be a good match for AMS for several reasons: it had several locations outside of the United States, it was close to my family, and the working environment was very relaxed—no more blue suits for me.

I was initially placed on a project based in Mexico City, but that project was winding down and after two months it was time to move on. I went to the staffing coordinator and expressed my interest in working abroad. This was on a Friday afternoon. On Monday morning I got a call that said everything was all set for me to go to Warsaw, Poland, where I would be working as a programmer analyst. They would need at least a one-year commitment. No problem! I trained in Denver for one month to learn the software, and three months after that, I was working overseas.

I cannot say enough good things about my experience in Poland. What began as a one-year assignment eventually turned into two years. I loved the experience of living with many nationalities. Learning about other cultures helped me grow as a person and taught me to be tolerant of differences. It was a challenge at first to learn which cultures start working at 8:00 A.M.

(Americans, Poles, and Germans) and which start working at 10:00 A.M. (Portuguese and Spanish).

After several years, I decided to leave AMS and started searching for a new position using an international-job search engine called jobserve (www.jobserve.com). Using a headhunter, I was hired as a consultant in Britain. I ended up returning to the United States as an expat in my own country, where I worked on a project, oddly enough, that was staffed with Norwegians and British. I never felt I was truly back in the States.

After the consulting stint was up, I wanted to return to Europe. I missed so many things about living abroad, like being able to get in a car or an airplane and in just a few hours be submerged in another culture and language. To remedy that, I phoned up my previous project manager with AMS and arranged to get rehired. Now I'm in Portugal on my latest adventure. Who knows what's next?

through the résumé-writing process. One of the best is *The Global Résumé and CV Guide* by Mary Anne Thompson (John Wiley & Sons, 2001).

- Network

  Networking is one of the single most important steps to creating opportunities in life, both career-related and personal. You'll be in the best position to find a job overseas if you take the initiative to network through your contacts. In fact, employers increasingly depend on job-applicant referrals from their own employees or other business contacts. So, if you are looking for a job overseas, start your network buzzing by telling everyone you know, and check out these two books.

  ★ **DYNAMITE NETWORKING FOR DYNAMITE JOBS: 101 INTERPERSONAL TELEPHONE AND ELECTRONIC TECHNIQUES FOR GETTING JOB LEADS, INTERVIEWS AND OFFERS**
  by Ronald R. Krannich and Caryl Rae Krannich (Contributor)
  (Impact Publications, 1996)

  This book is packed with practical information about how to get a job without applying for a position.

  ★ **A FOOT IN THE DOOR: NETWORKING YOUR WAY INTO THE HIDDEN JOB MARKET**
  by Katherine Hanson (Ten Speed Press, 2000)

  For job hunters who know they are supposed to network, but don't know how to get started, this book gives tips and advice for getting your foot in the door.

- Do Informational Interviews

  If you don't already have an established network, you need to build one. Start with the people you know—your brother, sister, friend, colleague, neighbor. Call 'em up and tell them you want to find a job in a global company. Ask them if they can refer you to any companies or friends who can help. Once you talk with one or two people and the ball gets rolling, networking becomes very easy. Most people love to help if they can.

After you've spoken to your personal contacts, branch out to professional colleagues or networking organizations like the chamber of commerce. With people you do not know well, ask for an informational interview. This is your opportunity to ask questions. Make it clear that you are not asking for a job, you simply want to learn more about their international work experience. (You want to pick their brain in a very diplomatic way.) It is helpful to prepare a list of questions ahead of time. You may even want to forward the list to the person you will be interviewing so they can be prepared. Conduct informational interviews with professionals in your field or people who have worked in your target destination. This is an opportunity to accumulate valuable contacts and to educate yourself.

Here are ten interview questions to get you started.

1. How did you find employment overseas?
2. How did you arrange your work permit?
3. What kind of education and skills are required for your position?
4. What is the normal salary range for your position overseas? (Name a specific destination, as salaries will vary by country.)
5. What types of things can be negotiated as part of an expatriate package? (Cross-cultural training, language courses, housing and cost of living allowances, paid trips home per year)
6. How is working abroad different from working in your own country? (Cultural differences, industry differences, tips on doing business abroad)
7. What kinds of company support can I expect to help me adjust to a new culture, new schools, and new life overseas?
8. How is an international experience viewed and valued by the home office?
9. What companies in your field send employees to work overseas?
10. Do you know anyone else with international experience with whom I could speak?

- Work the Net

You can network online through international communities, professional organizations, and message boards. On Monster.com's message board for "U.S. citizens who want to work abroad," international job seekers post messages asking advice from those who have worked overseas. You can get a wealth of insider tips on international companies, global positions, salaries, basic pointers on how to get started in the world of global work. Escape Artist, Expat Exchange, Expat Forum, and IAgora provide good networks to people who live and work abroad.

★ Escape Artist

www.escapeartist.com

Escape Artist is a site for American expats living overseas and offers resources dealing with relocation, tax concerns, and investing. Whether you need information on second passports, retiring overseas, or buying an island, Escape Artist is the place to start.

★ Expat Exchange

www.expatexchange.com

The Expat Exchange is the largest online community of English-speaking expatriates. Covering 140 countries, the information contained in the site covers all aspects, from your initial move abroad to your final return home. Among many other topics you will find clubs and organizations abroad, expat hangouts, relocation information, and networking boards.

★ Expat Forum

www.expatforum.com

The Expat Forum covers cost-of-living expenses, traits of successful expats, tips for making your assignment overseas a success, and a chat room with various forums about life abroad.

★ iAgora

www.iAgora.com

This site connects internationals worldwide and offers links to study, work, and travel abroad, plus tools, tips, and forums to share your experiences.

- Apply for International Jobs Online
  Use international job-search boards on the Internet, such as Monster.com, Jobpilot.com, or Overseasjobs.com, to search and apply for international positions. These sites have a range of resources and articles to help you find an international position, from résumé writing and interviewing to negotiating an expatriate salary and relocation package.

★ Idealist
www.idealist.org

Idealist specializes in jobs and internships in the nonprofit sector worldwide.

★ Jobpilot
www.jobpilot.com

Jobpilot is "Europe's career market on the Internet," with sites in fifteen European countries and Asia. In addition to posting your résumé and searching for jobs, you can read the Career Journal for tips on writing CVs, interviewing, negotiating, and moving.

★ Monster.com
http://international.monster.com

Monster.com is the biggest online job board, with extensive international resources covering sixteen countries. Go to the Work Abroad section to find worldwide job listings, apply online, take part in the message boards or chats, and read articles about work-abroad issues. The Global Gateway connects you to important resources for your move abroad. Simply enter the city where you live now and the country to which you are moving, and you will get information on visas, employment, employers, moving, and community contacts.

★ Overseas Jobs
www.aboutjobs.com or www.overseasjobs.com

In the Overseas Jobs link at Aboutjobs.com you will find jobs for students, recent graduates, expatriates, and adventure seekers, including a database for summer, intern, resort, and overseas jobs.

★ U.S. State Department

www.state.gov

This site has links to information on the Foreign Service exam, student programs, international organizations, travel warnings, passports, visas, and more.

- Use an Executive Search Company
  Executive search companies are usually focused on placing mid- or top-level managers and executives into key positions. If you are at this level or possess specialized technical skills, especially in the computer field, it is worth investigating what an executive search firm can offer you.

★ Executive Search.com

www.4executivesearch.com/executive.shtml

The site listed here will link you to a number of global executive search companies.

★ The Directory of Executive Recruiters 2001

by Llc Kennedy Information (compiler), Jennifer Shay, and Wayne E. Cooper (Kennedy Publications, 2000)

This extensive directory of thirteen thousand recruiters in sixty-one hundred locations in the United States can be searched by industry, functional expertise, geographic location, and individual specialty.

★ The Global 200 Executive Recruiters: An Essential Guide to the Best Recruiters in the United States, Europe, Asia, and Latin America

by Nancy Garrison Jenn (Jossey-Bass, 1998)

This guide contains background information on two hundred of the best-known executive recruiters, including their area of expertise and geographical orientation.

- Set a Time Line and Make a Plan B
  Decide how much time you want to spend actively pursuing a job that will send you overseas. Plan a trip to your target destination to gain a better understanding of the culture and job market. Once you have a time line in place, you'll be more motivated to accomplish your daily and weekly to-dos, bringing you that much closer to your goal of working abroad. You may want to have a back-up plan—doing a work-exchange program, moving overseas to conduct your job hunt, going back to school, or taking an international job in your home country—to increase your options.

- Research Salaries and Cost-of-Living Expenses
  Researching salaries in different countries can be a bit tricky. If you get an offer to work overseas, the offer could be considerably higher or lower than what you make in your home country. It can throw a lot of people off guard. However, the salaries in Europe are lower than in the United States in many professions. You'll have to weigh that potential con against the pros of increased vacation days, paid housing, and the opportunity to live in a foreign country. In certain cities it may appear that salaries are higher for some professions, but as is the case in the United States, you need to take into account the average costs of housing, groceries, and other expenses.

  Cost of living is easier than salary to calculate; however, that too depends on several assumptions, most notably the currency in which you will receive your salary. The Expat Forum site at www.expatforum.com has a function that calculates cost of living based on what a basket of goods would cost in different countries. Salary and cost-of-living adjustments are two of the many points you'll have to take into consideration when looking for a job and negotiating a relocation package.

- Negotiate Your Package

  When you finally get an appealing offer to work overseas, know how to negotiate your salary and expat deal. Write down the points of negotiation that you need to work out. Take into careful consideration the needs of your spouse, children, and any dependents, such as a parent. Try to persuade your company to arrange cross-cultural and language training for both you and the significant others who are joining you, and remember to discuss spousal job assistance, schooling for your children, and health-care arrangements for an elderly parent, if appropriate.

  In the excitement of preparing for your new life overseas, don't forget to inquire about repatriation assistance. Draw up an agreement explaining how your company plans to utilize your international professional expertise when you return to your home office. If you are planning to be away for several years, and especially if this is your first time abroad, you may need emotional and cultural adjustment support upon your return. Here are the most common negotiation points.

  - Salary (including currency in which your salary will be paid)
  - Duration of the contract
  - Signing and/or performance bonus
  - Relocation allowance (luggage, clothing, costs of refurbishing an in-country residence)
  - Moving expenses
  - Storage or shipment of your household belongings
  - Introductory trip to your host country
  - Orientation/adjustment assistance upon your arrival
  - Temporary living expenses before your departure (hotel, transportation)
  - Assistance renting or selling your current home
  - Paid in-country housing and expenses (utilities, maid service)
  - Number of trips home per year for the family, including your return tickets at the end of the assignment
  - Vacation days

- Health and personal insurance
- Language lessons for you and your family
- Cross-cultural training for you and your family
- Company car (and gasoline expenses) or driver service
- Reimbursement for forced sale or canceled lease on your car(s)
- Spousal job assistance
- Reimbursement for loss of spousal income
- Paid schooling for your children and reimbursement for school items
- Shipment of pets
- Tax assistance
- Health-care arrangements for an elderly parent
- Repatriation seminar for you and your family upon return home
- Temporary housing upon return home
- Shipment of goods to your home country and temporary storage, if necessary
- Relocation allowance upon return home

Monster.com, Jobpilot, and the Expat Exchange (see pages 266 and 267) all have resources that will help you negotiate the best deal. Following are a few other resources you might find helpful when moving abroad to work.

★ American International Schools

www.state.gov/www/about_state/schools

This is a listing from the U.S. State Department of American international schools. This is important information if you are relocating children and would like them to attend an American international school.

★ Expat Forum

www.expatforum.com

Expat Forum includes information on cost-of-living expenses, traits of successful expats, and tips for making your assignment overseas a success, and has a chat room with various forums about life abroad.

★ FIDI

www.fidi.com

If you want to make sure your breakables arrive unbroken, I recommend hiring one of the 730 approved national moving companies that belongs to FIDI.

★ Homefair

www2.homefair.com/calc/intsalcalc.html

To compare the cost of living in hundreds of cities worldwide, see Homefair's international salary calculator.

★ Monster Moving

www.monstermoving.com

Monster Moving has comprehensive information for your move overseas. Relocation essentials include tips on moving a family with teens, how to take care of personal paperwork like taxes and legal affairs, health matters, insurance, what to do with your home, choosing a child's school overseas, packing, transporting electronics, and shipping a motor vehicle.

★ Richard Lewis Communications

www.crossculture.com

Richard Lewis Communications provides excellent cross-cultural training and language skills development for individuals, professionals, and executives. The Cross-Cultural Assessor is one of the most advanced software tools for expatriates, human resources staff, and senior managers for assessing cultural types and learning how to communicate successfully across cultures.

★ Windham International

www.windhamint.com

Windham International offers cross-cultural training and resources to business professionals going overseas.

- Be Aware of Tax Laws

  As part of your negotiations, ask how your taxes will be handled. One very important tax law to be aware of when working overseas for more than 330 days is the Foreign Earned Income Exclusion. This is an annual tax allowance under U.S. tax law (under section 911 of the Internal Revenue Code) that allows Americans working abroad to exclude up to $80,000 USD for fiscal year 2002 (check with the IRS for subsequent years). You need to file Form 2555 with your 1040 Form to get the exclusion. Read about the details at www.irs.ustreas.gov/prod/forms_pubs/pubs/p54ch04.htm or order Publication 54 from the IRS. Your tax advisor should be able to show you how much money you can save by taking advantage of the Foreign Earned Income Exclusion. Before you go abroad, make sure you know how to deal with your taxes once tax season rolls around.

- Learn about Your New Country

  To make the most out of working in a foreign country, both professionally and personally, get to know your colleagues, embrace the differences, and enjoy the culture outside of the office. The more you know about the local customs, the better you will be able to do your job, earning yourself and your company respect.

- Take a Business Trip or Short-Term Assignment Abroad

  You may prefer to get your international work experience by going overseas on a business trip, taking part in a training course, or attending a meeting or conference. The advantage of a short-term experience like this is that you build professional qualifications and international exposure without the long-term separation from home, family, and friends. Keep in mind, though, that you will probably be holed up in a hotel and meetings for most of your stay, with relatively little time or opportunity to experience the local culture. Discuss a short-term option with your boss to find out how to be considered for such assignments, training courses, or conferences. Or, take the initiative to find out how attending an international conference could help your company and submit a proposal showing how you can add value to the company by attending the conference.

GO OVERSEAS AND GET A JOB

Believe it or not, recent grads, job changers, and adventure lovers may have a better chance of getting a job by just going overseas to look than they would by searching from their home country. Even MBAs who would rather be living abroad than waiting for the right opportunity to arise at home may prefer to pack up and go. It depends, of course, on your goals, your target destination, and whether or not you are supporting a family. The advantage of moving overseas to find a job is that success can come quicker than if you're searching from home. Although it may take some time to find the right job opportunity, you will have several months to acclimate yourself to the culture, the language, and the job market. You may also end up with much more responsibility than you could get in your home country. Usually you can stay in a foreign country for three to six months on a tourist visa—enough time to assess the work situation. Just make sure you bring enough cash to keep you afloat while you're looking, and be open to finding temporary work in the meantime.

When you do get an offer, your company may arrange your work permit or you may have to arrange your working papers in your home country. An alternative arrangement may be worked out in situations where no work permit is needed on a short-term or contract basis. You may end up with a local work contract and be compensated well by local standards, but you may not get the cushy expat package that so many international job seekers desire. However, your expenses may also be less and in the end you may end up saving more than you could at home. You will have to weigh all these factors when deciding how you want to get international work experience.

First things first: Decide on a target country, write your departure date on the calendar, and start planning. Make a month-by-month list of all the things you need to do before you go. (See chapter ten for planning guidelines and tips.) If you need to learn a new language, take classes before you go so you will know the basics once you arrive in the country.

Start saving your money and create a budget for the first few months (while you're looking for a job). Contact online communities to get insider information about what to expect from your target country's job market, housing situation, and cost of living. Get in touch with anyone you know in that country before you go so you have at least one "local" contact.

The next step is to go! Here are the things you'll have to do once you get there.

- Find Housing

  Once you arrive, find a place to stay by combing the classified ads in the local or English-language newspaper, by asking your contacts, or through postings at embassies, universities, and cafés. If you are traveling solo, you might want to find roommates who can help you with your job search and are willing to let you practice your language skills with them. Expats' Web sites like the Expat Exchange (see page 266) can put you in touch with the people and resources you need to make your move a breeze.

  To find international financial media, such as worldwide newspapers, databases, Web boards, and news, see www.qualisteam.com.

- Establish Your Network

  Once you are settled in your new home, your first visits should be to the local chamber of commerce and your embassy or consulate. These places provide information about international companies working in your city, organized business mixers and events, free seminars, and other activities you can join to meet people. Look into international clubs, such as the International Women's Club. And don't forget to access your online networks at Expat Exchange (see page 266), Escape Artist (see page 266), or iAgora (see page 266) to meet people living in your city.

  Here are a few more resources to help you with your move.

★ **EMBASSYWORLD.COM**

www.embassyworld.com/embassy/directory.htm

The site is an embassy and consulate search engine with worldwide links.

★ **FIRSTWORLDWIDE.COM**

www.firstworldwide.com

A Global Citizen's Perspective

Getting a Job in Poland

by Elizabeth Kruempelmann (ekruempe@hotmail.com)

Going back and forth between the States and some exciting destination across the big pond is part of building an international life. "Jumping around"—as some people refer to it—allowed me to gain certain skills stateside and then use them abroad, and vice versa, creatively weaving a unique international life and career path. After returning to the United States from Denmark and Germany, I was happy to be putting my house in order, saving money, paying off loans, building new marketing and sales skills, reconnecting with friends, continuing my language learning, and preparing for my next overseas sojourn.

After living and working near the Colorado Rockies for a year and a half, my German fiancé and I were off again. This time we were headed to Poland. Not knowing much about Poland, most people thought we were crazy. They pictured grumpy people in furry hats waiting in long lines to buy bread. Although breadlines were common during Poland's communist past, these images are relics of another era. The newly democratized Poland, not yet fully westernized when we moved there, was changing at a frantic pace. Big supermarkets with products from Western countries were springing up every few months, multinational businesses were descending on the big cities, and life was improving for many people, especially the younger generation, which was eager to be part of a westernized world. Gray images of Warsaw portrayed in the press were replaced by dazzling neon signs advertising Marlboro cigarettes, Coca-Cola, and American's favorite fast-food joints: McDonald's, Burger King, and Pizza Hut. Poland was an exciting and happening place to be.

In contrast to Germany, where a few years earlier I struggled to find any kind of work, Poland proved to be a dynamic and hot job market. As I met and networked with members of the international community in Warsaw, I ended up with three job offers in the first three weeks in my field of sales and marketing. I accepted an offer to work as a partner and marketing manager at a small advertising start-up run by expats. I was technically hired as an American marketing consultant. My partners assured me that they could justify my employment by explaining that they needed a native English speaker trained in Western-style sales and management to deal with our multinational clients. These were skills that few Polish locals had. The work permit situation in Poland at the time was terribly bureaucratic and many professionals were working there on legally fuzzy, yet locally accepted terms (see more about work permits on page 220).

My job turned out to be incredibly interesting since the advertising industry in Poland was booming. The skills I had gained in the States were used to sell advertising methods that had never been tried in Poland, from coupon advertising to direct mail. Under communism there was little need for advertising and marketing, so there were few precedents for what kinds of marketing would work best. Luckily, the Poles were relatively open to new and untried methods. The experience I gained traveling to many of the major cities in Poland to present our services to international and Polish companies was incredible. Knowledge of Polish was not necessary, as most employees of international companies were required to speak and conduct business in English. In addition, as a working partner, I was involved in the strategic decision making of a start-up business, an opportunity that would have been harder to come by in the United States. Attending business events was essential to building business contacts and personal friendships. The expatriate community in Warsaw was relatively small and often congregated at mixers organized by the American or British chambers of commerce, the press, and large companies.

Although salaries in Poland were quite low at the time, I earned two to three times the average salary, which was enough to live on. I didn't get rich, but the experience made up for it in full.

Read more about Elizabeth Kruempelmann's international experiences in chapters two, five, and seven.

This is an excellent online connection to chamber of commerce offices worldwide that can provide you with information about local business climates.

- Learn the Language

Among the things you should do sooner rather than later is get signed up for a language class, if necessary. If you wait too long, you'll get too much in the habit of speaking English.

University courses are cheap and allow you to meet locals. You could also arrange a language exchange with a local student who gives you an hour of language lessons in exchange for an hour of English lessons. This is a free way to learn and an easy way to make a friend.

★ **BERLITZ (SEE PAGE 102)**
www.berlitz.com

★ **INLINGUA**
www.inlingua.com

This is a reputable language school with locations worldwide. There are over three hundred InLingua language schools that offer language and cross-cultural courses (and jobs for English teachers).

- Set Up Informational Interviews

As you start to attend business events and get to know people, set up informational meetings to learn about local companies or industries. See the networking tips section on page 264 for some good questions to ask. Remember that an informational interview is an opportunity to gather information. Your contact may be offended if you come right out and ask for a job. However, do let people within your network know what you have to offer and that you are looking for specific employment. Work your network every day until you get an offer.

An excellent site that gives you the steps to the informational interviewing process is http://danenet.wicip.org/jets/jet-9407-p.html.

- Take Advantage of Internet Opportunities
  Don't forget to check out Internet opportunities! Read the local newspapers and watch the local news to find out which Internet companies are expanding. Maybe they need your expertise to finally reach markets in your home country.

- Stay Involved in the Local Culture
  Make sure you do whatever is necessary to stay involved in your local culture. This can mean getting a job at a local restaurant to pay the bills, arranging your own internship, volunteering, or filling your days with networking meetings. Make sure that any part-time jobs you take can provide you with enough money to live on if finding your dream job takes longer than expected. Don't get too stressed out about the amount of time it takes to find a job because employers will sense your frustration. Simply relax, stay persistent, and soak up the experience of living in a foreign country.

- Apply for Jobs
  While you are getting to know people and inquiring about jobs in your field, you'll want to simultaneously apply for jobs through local channels, like job-hunting Web sites, newspaper advertisements, and job fairs. Monster.com is one of the biggest online job boards with international job listings. See page 267 for more information. If you can, find an online version of the country's main newspapers and apply for a few jobs before you leave home. If a company knows you will be available for a face-to-face interview, they're more likely to consider your application. Once you are in the country, find out about job fairs and local recruiting events through newspaper and radio advertisements and the local university's career office. It is always wise to have a local friend help you out with the cover letter and application process. Find out if it is common practice to follow up your application with a phone call, or to write a thank-you note after an interview. Although these are common practices in the United States, other countries may have their own procedures.

CREATE YOUR OWN OPPORTUNITY ABROAD

Picture this: You are in a foreign country trying to find a job. You don't have a work permit and the job prospects are sparse. What do you do?

Start thinking like a business owner rather than a job hunter. Set up your own consulting or freelance business and start to scout for clients, not a job. Send out brilliant proposals instead of résumés. Have business meetings with potential clients instead of interviews with potential employers. Propose clever ideas for improving your client's business. And when the time is right and your client is ready, name your price. They can accept it or reject it, but eventually you will probably end up negotiating the terms, just like you would when accepting a job offer.

Setting up your own business is not as hard as it sounds. It is essentially job hunting in disguise. And, it is a little-known trick to getting around the work-permit issue . . . at least for a one-person business in the short run. (If you want to set yourself up as a corporation with more employees than just yourself, the process becomes more complicated.)

To develop your own business you have to sell yourself as a specialist. This is true for traditional job hunting as well as entrepreneurial pursuits. All of us have unique interests, talents, and skills. The problem is that most of us take our innate interests for granted and don't even consider that they could become money-making opportunities at home or abroad. Think about what you like to do. Think about the skills you like to use. And then combine your interests and skills into a product or service that you can market to a target group of people. This is your specialty, and your specialty is what you will market and sell.

For instance, because you are reading this book I assume you have an interest in traveling. What skills do you use when you're seeing the world? If you enjoy taking photos of your favorite travel destinations, perhaps you could make photography your specialty. Frame your photographs and market them to international hotel chains, restaurants, or businesses. Maybe you have language skills that you could turn into a translation business or writing skills that the local English-language newspaper could use. Help a new Internet company expand to your home-country's market, or use your Web-development skills to create Web sites for businesses.

Instead of developing your résumé, develop a marketing brochure. Approach small companies with a proposal explaining what you can do for them. Once you take ownership of your talents, thinking like a consultant, freelancer, or entrepreneur becomes easy.

If you are a one-person show, you simply have to know what you can offer and start pitching it to your target market. Depending on your product and target market, you might have to spend some time analyzing the market for your product, the competition, and the costs of operating your business. You can create your business informally by having your salary deposited into an account, usually in your home country, that has your name or business name on it. In some countries you may also need a tax identification number obtained from the local tax office or administrative agency responsible for issuing tax identification numbers, since you may be legally responsible for paying taxes after a certain amount of time working there. Have some professional business cards printed, and create a Web site to promote your business.

- A Foot in the Door

  By setting yourself up as a consultant or freelancer, you may also be able to sell yourself to a company. Instead of hiring you as a full-time employee, a company can hire you on a contract basis, which may be an advantage for both of you. When I lived in Poland, the small advertising company I worked for wanted me to work full-time. Because it was complicated to get a work permit, we agreed that they would put me on their payroll as a full-time marketing consultant and I would issue them an invoice with my American company address on it every month. Setting myself up in my own marketing consulting business was as simple as that. Additionally, in many countries the company might be able to avoid paying expensive social security and other taxes by hiring you as a consultant. This is a good negotiating point when you are trying to convince a company to hire you as a consultant. To avoid oversimplifying a potentially more complicated process, let me just add that this type of working arrangement was allowable in Poland at that particular point in time. Country laws regarding work issues can change with new governments and the ensuing economic, political, and social developments, so be sure to do your research before you start working.

Consulting or working on a freelance basis is also a good way to prove yourself. In the future you could be considered for a long-term position if one becomes available. If a company really wants to hire you as a full-time employee after you have proven yourself, they'll find a way to officially explain why you must be hired over a local.

- A Laptop Equals a Portable Job
  If you move from country to country every few months to every few years (like my husband and I), you want to be able to work from where you live. You can do lots of interesting work from a laptop, including writing, Web-page development, graphic design, software design, content development, research, message board hosting, or other Internet-related jobs. In the last ten years, I haven't lived more than two straight years in the same place. Although I've been lucky enough to find fulfilling work in all the countries we have lived in, I'm just now learning to blend my personal travel aspirations with my writing, message board hosting, and cross-cultural consulting projects—professional work I can do from anywhere, anytime.

- Talk with Other Entrepreneurial Types
  Find out how other people are living their lives overseas. If you hear ideas that strike your fancy, make contact with the folks and ask them for advice. One middle-aged American couple I heard about wanted to move to France. Now they buy and restore old farmhouses and rent them out to tourists. Another American living abroad gives seminars and workshops in photography and art. And one woman who lives abroad permanently uses her graphic skills to design newsletters for clients in the United States. These people are living where they want and the way they want. And with a little ingenuity, you can too.

★ **Business Culture**
www.businessculture.com

Business Culture sells reports on doing business in other countries, including etiquette and cross-cultural communication issues.

★ Global Writers' Ink

www.inkspot.com/global

Global Writers' Ink covers all aspects of international writing and marketing, including how to develop ideas for the international marketplace and how to survive as a writer abroad.

★ World Franchise Consultants (WFC)

Wfcnet@cris.com

WFC assists entrepreneurs in finding the most suitable business or franchise opportunity, including international franchising.

★ Anatomy of a Business Plan: A Step-by-Step Guide to Starting Smart, Building the Business and Securing Your Company's Future

by Linda Pinson and Jerry Jinnett (Dearborn Trade, 1999)

This book shows you how to create a polished business plan, access financing from lenders, and market your services using the latest Web-marketing strategies.

★ Start Your Own Import/Export Business

edited by Joann Padgett (Prentice Hall Press, 1994)

This book explains different approaches to the world of import/export, from working as a consultant or middleman to selling to wholesalers in another country.

See also EmbassyWorld.com (page 275) and FirstWorldwide.com (page 275) for information on embassies, consulates, and chambers of commerce.

GET AN INTERNATIONAL JOB AT HOME

If you are not yet ready or able to go abroad, or you've just returned from an overseas stint, but still want to work with foreigners, use your language skills, get involved with international causes, or work in an international capacity, it is possible to get international exposure without leaving your home country.

A GLOBAL CITIZEN'S PERSPECTIVE

Creating Global Work That Works for You

by Eve Berton, EveArt (eberton@it.com.pl or www.lesgalleries.com)

I had always wanted to live in Europe, so when my boyfriend got a job in Warsaw, Poland, and asked me to go along, I walked away from the sweet security of a $70,000 USD-a-year paycheck. I didn't bother to line up a job, and this failure to plan landed me in a whole lot of trouble. But it also introduced me to a new life that brings more reward, challenge, freedom, and adventure than anything I could have planned.

While I took a considerable risk in moving abroad, I had ample savings should things turn sour. And they did. The relationship crashed and I moved out on my own, knowing just enough Polish to get around. Rather than look for a full-time job in public relations or writing or publishing (my professional skills) I put myself at the mercy of the freelance market. I made contacts through the expat community and volunteered for Polish organizations. I did whatever was needed: public relations training, grant writing, organizational development consulting, newsletter development, media relations, job-search counseling, travel writing.

I lived frugally, traveled extensively, and dragged a sketchbook wherever I went. I printed some of my sketches as greeting cards and marketed them to major hotels. As an artist, I discovered it was legal to sell works of art—that is, pieces inspired by my creative urge—without a company, and without a work permit. Because I also intended to sell commissioned and printed pieces, I decided to start my own business. People began to commission drawings and sketches, and I found my creative muse. By this time my Polish was strong enough to get me through any business transaction, so I formed my own company to sell my art. (It took the patience of Job and a will of steel to master the Kafkaesque tax bureaucracy.) I work out of my flat, which is also my art gallery, have several lines of cards in production, sell paintings and cartoon tapestries (see www.berton.ws and www.lesgalleries.com), and am illustrating a children's book.

Money is not guaranteed, and life is a bit less secure than I was used to in the States, but I am living the life I always dreamed of. I work according to my own schedule, travel when I have the need, and live creatively. I don't support a car, my flat is furnished with repainted discards, and I dress chic in a little simple black. I didn't follow any path, just the scent of life.

Before I left the States, I couldn't imagine living outside the maximum security realm of paycheck, credit purchases, and unemployment insurance. There's nothing wrong with that, but "maximum security" also means the tightest prison.

If you want change: take risks.

If you want security: plan your jump.

If you want adventure: allow for the unplanned.

If you want to write a book: live something worth writing about.

If you think it will be easy: think again.

My first job out of college was a marketing position in a prestigious think tank in Washington, D.C., whose renowned economists researched and published books on international economics. Although my position was not what I would consider international, the environment in which I worked was a mini–United Nations. My colleagues, from whom I learned a tremendous amount, were from Taiwan, Bangladesh, Colombia, and Brazil, and the topics we dealt with were global in nature. This job stimulated my curiosity about the world and influenced my decision to return overseas to pursue an international life.

You can easily start or continue to develop your global skills by working at an international organization. Major cities such as Washington, D.C., New York, and San Francisco house many organizations doing business overseas in law, banking, politics, education, and the nonprofit sector. International jobs might be harder to come by in smaller towns or cities. Check with your local chamber of commerce office to find out about global companies in your town.

There are also opportunities for you to apply your international work experience to jobs in your home country. A friend of mine who was a Peace Corps volunteer teacher in Guatemala, returned to her hometown to teach bilingual high school students. Her international and language expertise paid off and led her to a rewarding career.

These are the points you need to consider if you want to get an international job at home.

- Decide What "International" Means to You
  We all have different concepts of what an international job is. It could involve topics of a global nature, such as international economics or global environmental concerns. It could mean working with clients from around the world on a

global campaign. It could require you to use your language skills. Or it could mean dealing with people from different cultures on a daily basis. From advertising agencies to administrative positions and international exchange organizations or language schools, you can find jobs with an international element in almost every profession. Not every international job will offer all of these elements, so determine your priorities before you start your job search.

- Start Networking
  Start looking for an international job in the same way you would a local job, using your contacts to generate leads. Focus on international organizations and opportunities. If your long-term plan is to eventually go abroad or be sent abroad, pick your companies and opportunities with that in mind.

- Learn a Foreign Language
  To make yourself more marketable to international organizations, learn to speak another language proficiently and get some international internship, volunteer, or professional experience. If you don't have either of these, you still might be able to get into an international organization at some level and work your way up. Evaluate the skills, personal characteristics, and abilities you have that will help you fit into the company of your choice.

- Get Involved with the International Scene
  Read international publications, build a network of international friends, and take language classes, all the while keeping an ear open for job opportunities with a global bent.

- Stay Open to Other Ideas
  While you are searching for an international job at home, keep other internationally-oriented ideas in mind, such as volunteering to teach English to immigrants in your city or taking part in an educational tour of a country you might like to work in someday. As you read through this book, jot down other ideas that strike you.

A Global Citizen's Perspective

Applying Peace Corps Experience at Home

by Michele M. Carlson (missmich27@hotmail.com)

I am a high school history teacher at a public school in Chicago, Illinois. My career choice came as something of a surprise to everyone who knew that my dream had been to work overseas. I have always loved anything international. Foreign languages intrigue me, I loved my international relations classes in college, and I jumped at the chance to study in Denmark and do an internship in Brussels during my third year at university. Upon graduation, I wanted to work for an organization that focuses on international development. So I joined the Peace Corps, hoping to get practical experience for my future career, and at the same time save the world. During the two years that I spent in Guatemala working as a rural youth educator, I realized that what I really wanted to do was become a high school history teacher in my hometown.

The Spanish fluency I gained in Guatemala definitely came in handy as I looked for a job. I was hired by Kelly High School as a member of their bilingual and history departments. Although the population of Kelly is predominantly Mexican, only a small percentage of the students qualify for the bilingual program. Because I speak Spanish, I am able to communicate beyond simple exchanges with the students. I hold conversations with parents about their child's behavior, and I converse with students who prefer speaking Spanish even though they are fluent in English. I understand jokes and double entendres, not because I know the words, but because I understand the context. Because I speak Spanish, parents have an outlet through which they can voice their concerns, and because I lived in a Hispanic culture, students have a teacher with whom they feel at home.

I am lucky to have had classes full of motivated and interested students who want to participate and learn. I am able to share my own experiences of living and traveling abroad to create lessons that are current and interesting. Recently my seniors completed a unit on Liberia and Sierra Leone during which we studied the use and exploitation of child soldiers. This unit was inspired by a Liberian man I met who had been imprisoned during his country's civil war for speaking out against the government. The final assessment for the unit was not a test, but instead participation in an Amnesty International letter-writing campaign to Charles Taylor, president of Liberia, to urge him to convince the Revolutionary United Front of Sierra Leone to stop using children in war.

Wanting to actually experience world cultures, my current students decided that we need to travel overseas as a group. I am pleased that they want to travel somewhere to make a difference, for example, to participate in a Habitat for Humanity–sponsored event.

While I still dream of living and working in a country other than the United States, for the moment I enjoy teaching my bunch of inner-city high school students, and through them, indulging my international interests. I am able to study and analyze events around the world, and I am able to travel as a complement to the classes I teach. I use Spanish on a daily basis, and I take courses that enhance my teaching. Most importantly, I have been able to pass my love for things international on to many of my students, who are beginning to see that there is life beyond the U.S. and Mexican borders.

SHOULD YOU STAY OR SHOULD YOU GO?

Now that you've read about the variety of ways you can get a job overseas, you might be trying to decide between staying in your home country and getting an international degree or a position with an international company, or packing your bags and trying your luck abroad. The answer depends on your career goals and your determination to live overseas as soon as possible.

These different options for working abroad and finding international work opportunities should help you sort out your international career direction and plan the next steps to your dream life abroad.

My personal experience tells me that if you have a burning, passionate, and immediate desire to live overseas, if your heart races every time you hear about someone else's exciting adventures abroad, you simply have to plan wisely and go for it—with or without a job in hand. This approach is for confident, adventurous risk takers who are prepared to do what it takes to make their experiences abroad the best they can be. This approach is not for everyone, so make sure you are the type of person who is willing and able to make the most of it.

Get yourself overseas in some capacity, whether it is through an organized work or volunteer program, a language course, extended travel, a visit to a friend, or some other way. Once you are there you will be in a much better position to network, check out the work and living situations, and evaluate your prospects more realistically. If you are having problems getting a work permit, check with other expats to see how they arranged theirs. You may be surprised to discover legal loopholes or companies willing to hire you and arrange your working papers. Or you may indeed discover that work permits and jobs are few and far between, in which case teaching English, freelancing, or working in the service industry may be the only ways you can earn money, stay in the country, and gain valuable professional skills. But remember, if things don't turn out the way you hoped, make the most of what you have accomplished. You can always go home, or try the international job market again in the future.

If you are worried about how going abroad without a job will reflect on your reputation and career prospects, you're not alone. Not everyone would be happy diving into the unknown without a life jacket. And there are some who will see your ambition and risk-taking efforts as directionless diversions from your career path if you don't market your experiences creatively. On the other hand, there are employers who value international experience and will recognize that your unique character and skills are assets to their company. After all, you will have shown that you are willing to take the initiative to get international business and life skills to better prepare you for an international career.

If you don't want to pack up and move overseas without a job in hand (which truth be told, can be a real pain), you may want to build up your skill base at home first. This will give you time to network both within your company and with outside contacts until the right job opportunity presents itself. In the long run, establishing a professional track record at home may situate you more successfully for an international expatriate position down the road. While you are building your career, get the international job search started since it can take some time. Just make sure you position yourself strategically from the start, and stay focused on your goals. Don't let what sounds like a great career move distract you away from your dream of working overseas.

INDUSTRY LINKS TO INTERNATIONAL JOBS

Regardless of whether you head overseas right away or start building your global career at home—you will be acquiring the skills that are key to creating international life and career. To help you research international opportunities in business, government, and law, as well as international agencies and nonprofit organizations, this section contains a list of umbrella Web sites that list related industry and job links and specific companies that offer international job opportunities to qualified applicants.

ADVERTISING

★ **ADVERTISING WORLD**

http://advertising.utexas.edu/world

For advertising and marketing students and professionals. You will find links to advertising-related resources.

 **YOUNG & RUBICAM, INC.**

www.yr.com/companies/careers

With offices worldwide, global accounts, and overseas positions.

BANKING

BANKWEB

www.bankweb.com

An umbrella site that offers a comprehensive list of domestic and international banking sites.

 **CITICORP**

www.citicorp.com

Offers global job opportunities for new graduates and professionals in the career section of the site.

BUSINESS

YAHOO

www.yahoo.com

Along with other search engines, Yahoo is a good place to get background information on specific companies.

 **TEXACO**

www.texaco.com

Has offices worldwide and offers an Executive Business Analyst Program that develops leaders through domestic and international business rotations.

CONSULTING

MANAGEMENT CONSULTING NETWORK INTERNATIONAL (MCNI)

www.mcni.com/onassociations

Includes information about the International Council of Management Consulting Institutes (ICMCI), whose member institutes worldwide certify management consultants with the CMC (Certified Management Consultant) title.

Cross-Cultural Services

★ International Human Resources and International Relocation Management

Society for Human Resource Management (SHRM) Forum

www.shrmglobal.org

A membership society of global human-resource (HR) professionals who share information and network. View its global HR job listings online.

★ Berlitz Cross-Cultural Services

www.berlitz.com

In addition to being the foremost language trainers, Berlitz has jobs available for cross-cultural trainers, resource coordinators, and positions in the related field of international relocation. Internships are also available.

Government

★ The Agency for International Development (USAID)

www.usaid.gov/about/employment

Lists international job openings.

★ The Central Intelligence Agency (CIA)

www.odci.gov/cia

Has job listings and contact information for those interested in becoming a CIA agent or analyzing intelligence information.

★ Defense Intelligence Agency (DIA)

www.dia.mil

Offers only a limited number of support assistant positions in U.S. Defense attaché offices at embassies worldwide.

U.S. Office of Personnel Management

www.usajobs.opm.gov

The main government job bank, but you will have to sort through the jobs to find the international ones that fit your interests.

The U.S. State Department

www.state.gov

Has a Web site with links to information on the Foreign Service exam and international vacancy announcements for the UN and other international organizations.

Health

Doctors Without Borders

www.doctorswithoutborders.org

Delivers emergency medical care to communities and victims of armed conflicts, epidemics, natural disasters, and to others who lack decent health care due to geographic or social isolation.

International Medical Corps

www.imc-la.com

A global humanitarian nonprofit organization whose mission is to provide health-care training and relief programs to diminish suffering and save lives throughout the world. Offers paid positions worldwide.

International Organizations

The International Institute for Sustainable Development

http://iisd1.iisd.ca/ic/sb/direct/sdun.htm

Has links to UN organizations, banks, and overseas development assistance agencies.

United Nations

www.un.org/depts/ohrm

The United Nations (UN) and its thirty affiliated organizations work to promote world peace, justice, human rights, international law, humanitarian assistance and development. Nearly every country belongs to the UN. See the Web site for more details on job opportunities with the UN and its family of diverse organizations.

The World Bank

www.worldbank.org

The World Bank is devoted to eliminating world poverty. Although the hiring process is extremely competitive and most new hires are midcareer professionals, students and MBA grads may be able to get an internship, take part in the young professionals program, or obtain a scholarship.

Internet Companies

Look up your favorite Internet sites to see if they have international positions or are expanding overseas. Try www.amazon.com, www.yahoo.com, or www.ebay.com. Also research foreign Internet start-ups that are expanding to your country. You may be able to work from home from any location in the world.

Journalism

Cable News Network (CNN)

www.cnn.com/jobs

Has international jobs and internships posted on their site.

National Association of Broadcasters Employment Clearinghouse

www.nab.org/bcc/default.asp

Has a career center on its Web site with job postings in broadcasting.

Reuters

www.reuters.com/careers

"The world's leading financial media and professional information company" offers opportunities for travel, international assignments, and graduate training programs.

Sun Oasis Jobs

www.sunoasis.com/global

Has links for searching worldwide jobs for writers, editors, and copywriters.

Law

American Bar Association

www.abanet.org

Provides links to career counseling and international opportunities.

Baker & McKenzie

www.bakerinfo.com

A renowned international law firm with international jobs listed on their site.

Nonprofits

Greenpeace

www.greenpeace.org/information.shtml

A nonprofit organization whose efforts to save the environment are known worldwide.

InterAction/American Council for Voluntary International Action

www.interaction.org

A nonprofit umbrella organization of over 165 nongovernmental organizations working for sustainable development, refugee and disaster assistance, and humanitarian aid.

Idealist

www.idealist.org/career.html

An excellent job-search database that lists every nonprofit job site or directory that they could find on the Web.

★ International Red Cross

www.redcross.org/jobs

A nonprofit organization with a particular emphasis on helping disaster victims worldwide.

★ Nonprofit Job Source

www.jobsourcenetwork.com/nonpro.html

A highly recommended umbrella site with links to hundreds of nonprofit sites worldwide where you can search for full-time, part-time, internship, and temporary jobs and careers with all types of nonprofit organizations. The job listings are connected with the Idealist Web site shown on page 181.

Other Opportunities

★ American Society of Travel Agents (ASTA)

www.astanet.com

The world's largest association of travel professionals. Their Web site includes information about job fairs and links to ASTA travel schools.

★ Association of International Educators

www.nafsa.org

1307 New York Avenue, NW
Washington, DC 2005
Phone: 202-737-3699
Fax: 202-737-3657

NAFSA promotes the exchange of students and scholars to and from the United States. NAFSA members believe that international educational exchange promotes learning and scholarship, builds respect among different cultures and nations, and encourages leadership in an increasingly global world. NAFSA is a membership organization for professionals in the field of international education at the postsecondary level. NAFSA offers its members professional development programs, travel grants, and educational opportunities.

★ **Berlitz**

www.berlitz.com/franchising

If setting up a language school in a foreign country sounds like a great way for you to have your own business, then a Berlitz franchise might be for you.

★ **Institute of Certified Travel Agents**

www.icta.com

Offers one of the most respected training programs in the travel industry to become a Certified Travel Counselor (CTC). You can research and plan your travel career on their Web site.

★ **International Association of Culinary Professionals**

www.iacp.com

A nonprofit association for culinary professionals from over thirty-five countries involved in culinary education and communication or the preparation of food and drink. It provides continuing education and development for its members and has a job bank of worldwide positions.

★ **Society of American Travel Writers (SATW)**

www.satw.org

This is *the* organization to belong to if you are a travel writer, although you must be sponsored for membership. Members are considered "legitimate" travel writers and are respected by the travel industry. To learn more about the benefits of membership, log on to the Web site.

★ TERAGLYPH

www.teraglyph.com

An atlas of franchise opportunities in over four thousand sectors that you can search alphabetically or by category.

★ VANCE INTERNATIONAL SECURITY SYSTEMS

www.vancesecurity.com

Vance specializes in protection services with personnel, contacts, and offices in key locations throughout the world. Their services include executive protection, uniformed security officers, investigations, crisis management, asset protection, temporary labor, and security protection. Vance's security practitioners are trained in U.S. Secret Service doctrine.

PART THREE

Making Your Global Dream a Reality

Practical Stuff You'll Need to Know

The best way to predict the future is to create it.

UNKNOWN

This part of the book will show you how to make your global dreams a reality, from raising the funds to travel, preparing to depart, maximizing your time abroad, and even returning home with ease. Then it will be up to you to leave your country and all of its conveniences, and the security of friends, family, and a familiar routine, for the sake of reaching new plateaus of self-enrichment.

CHAPTER NINE

Affording It: Funding Your Sojourns

Where the determination is,
the way can be found.
GEORGE S. CLASON

There is a popular myth that says going abroad has to be an expensive undertaking. It's true that you could easily spend $15,000 USD to study in Australia for a year or drop $3,000 USD to take part in a three-week community service volunteer program in Africa. But there are also many ways to fund your sojourns without breaking the bank or going into major debt. In fact, people often pay more to go abroad—whether to travel, study, volunteer, or work—than they have to.

I wish I had thought about financial planning before I (meaning my generous parents and my bank) dished out quite a big chunk of cash to study overseas. I realize now that I would have qualified for a variety of scholarships and travel grants that would have paid the total costs of my study-abroad program. Unfortunately, I didn't investigate any funding ideas, assuming incorrectly that I wouldn't qualify. The reality is that many funding opportunities like scholarships, work/study programs, travel discounts, and fund-raising efforts aren't pursued by qualified students, professionals, and adventurers. That leaves money on the table for candidates like you.

The less debt you have to pay off in the long run, the more freedom you will have to pursue your interests and save for the future. The main reason I returned to the United States after my second overseas trip was to get a higher paying job to pay off my college and study-abroad loans. If I hadn't had these debts looming over me, I would have gladly stayed abroad. On the positive side, student loans did make it possible for me to study abroad and I was able to defer my payments for a while. Loans can be a good option for covering those expenses other sources won't cover.

Don't let what, in some cases, may seem like extraordinary costs prevent you from pursuing your overseas dreams. Many people find creative ways to finance their global ambitions and you will too. Affording an international education is a life choice. Consider it an important investment in your future. The benefits will be significant, measurable, and absolutely worth it.

Before reviewing the funding opportunities available, let's take a minute to inventory the types of expenses you'll have.

- Transportation costs
  You need to get to where you are going, right? That usually involves a plane ticket at the very least. If you're traveling to a remote location, your transportation costs will not be cheap.

- Program fees
  If you are taking part in a study-abroad, intern, volunteer, professional, or other type of exchange program, you'll have to pay program fees, like registration and placement costs.

- Tuition
  Similar to colleges at home, students have to pay tuition to study abroad. Sometimes tuition costs the same as the student's home institution, and other times students are charged the overseas university's tuition rates. Rising tuition costs are often the most common reason students back away from pursuing opportunities overseas.

- Expenses
  Although some internship and volunteer programs are all-expenses paid, many are not. Therefore, you'll be responsible for covering rent, transportation, food, phone bills, insurance, utilities, and all the other unexpected items that seem to crop up.

- Personal travel and entertainment
  Many times the funds you earn or raise can be used at your discretion, meaning you can spend them on whatever you want, including personal travel, movies, museums, or perhaps a gorgeous pair of Italian leather boots.

Scholarships

Scholarships are often thought to be reserved for only the best and brightest students or those who can prove financial need. People don't realize there are many less restrictive scholarship opportunities available from a large range of sources. Students can approach their university for grants or loans, and organizations like Fulbright and Rotary provide extensive funding, not only for students or scholars, but for professionals as well. And even the government has scholarship programs just waiting for someone like you to apply.

UNIVERSITY SCHOLARSHIPS AND LOANS

Every college and university has a financial-aid office that provides information about scholarships, loans, and work/study programs. If you already receive financial aid, you will probably be able to apply it to a study-abroad program. If you do not currently receive financial aid, you still may qualify for scholarships designated specifically for studying abroad.

EXCHANGE PROGRAM SCHOLARSHIPS AND GRANTS

Many study-abroad and exchange programs offer scholarships of varying amounts to qualified students. Some are based strictly on financial need, while others are based on fields of study or even more general terms. Check with your exchange program and its affiliates to see if you qualify for any of these options, or contact the following resources.

American Institute for Foreign Study (AIFS)

www.aifs.com/college

River Plaza
9 West Broad Street
Stamford, CT 06902
Phone: 800-727-AIFS

Each year AIFS awards over $230,000 USD in international scholarship funds, including fifty study-abroad scholarships of $1,000 USD each for the spring and fall semesters, and fifty summer scholarships of $750 USD. Other funding opportunities include minority scholarships, the AIFS Foundation Kraków Scholarship, the Paul Moonves Memorial Scholarship, and the Tcherepnine Scholarship. Application deadlines are October 15 for the spring semester and April 15 for the fall semester.

Council on International Educational Exchange (CIEE)

www.ciee.org/study/scholarships

205 E. 42nd Street
New York, NY 10017
Toll-free: 888-268-6245
Phone: 212-822-2600
Fax: 212-822-2699

CIEE's International Study Programs (ISP) offers six types of scholarship aid:

- The Bowman Travel Grant
  Issued by CIEE/Council Study Abroad, the Bowman travel grant is awarded to an American high school or undergraduate student who wants to study, work, or volunteer in a nontraditional country. Countries excluded from the grant are the United States, Australia, New Zealand, Canada, and many European countries. The amount of the travel award varies based on the destination. The application deadline is April 1 for summer, fall, and academic year programs, and November 1 for spring semester programs.

- CIEE-International Study Program (ISP) Scholarships
  CIEE offers ISP (International Study Program) scholarships of $500 to $1,000 USD each year to Council Study Center applicants who show academic excellence and financial need. To qualify, students must be attending Academic Consortium member institutions (all of which are listed on the CIEE Web site).

- Contemporary Cuban Society Program
  A limited number of $500 USD scholarships are available to undergraduates and graduates for participation in the summer Contemporary Cuban Society Program.

- Department of Education Scholarship for Programs in China
  This is for U.S. juniors, seniors, or graduate students pursuing teaching careers in modern foreign languages or area studies, who have completed two years of college-level Mandarin Chinese, who can demonstrate an interest in pursuing postgraduate work related to Chinese studies, and who can prove financial need.

- Jennifer Ritzmann Scholarship for Study in Costa Rica
  This is a $1,000 USD scholarship that must be applied toward program fees for participation in the Tropical Biology Program at Monteverde, Costa Rica.

- Robert B. Bailey III Scholarship
  This scholarship, usually in the amount of $500 USD to be applied toward the applicant's program fee, is awarded to American students who belong to an underrepresented ethnic group and who apply to CIEE's Council Study Center program or a university Direct Enrollment Service program for a summer, semester, or year.

PRIVATE SCHOLARSHIP FUNDS

Scholarship money to study and work abroad can be obtained from private funding sources, the biggest of which is Rotary International. Private funds can also come from private citizens, corporations, and organizations who donate money.

★ **ROTARY INTERNATIONAL**
www.rotary.org/foundation/educational/amb_scho/index.html

One Rotary Center
1560 Sherman Avenue
Evanston, IL 60201
Phone: 847-866-3000
Fax: 847-328-8554

The Rotary Foundation is the world's largest privately funded international scholarship program. Their aim is to promote international understanding by sponsoring Rotary scholars throughout the world. Students, graduates, and professionals pursuing vocational studies can apply for scholarships at local Rotary clubs; however, not all Rotary clubs offer the full menu of scholarships and some Rotary clubs offer none at all. Rotary awarded more than twelve hundred scholarships in 2000–01. Recipients from sixty-nine countries studied in sixty-four different countries through grants totaling twenty-six million dollars. The Ambassadorial Scholarship Program, with three types of scholarships, is the world's largest privately funded scholarship program. All applicants must have already completed at least two years of college-level course work or have an equivalent amount of professional experience. See the Rotary Web site for other eligibility criteria, which vary per scholarship.

There are three types of Rotary scholarships available: Academic-Year Ambassadorial Scholarships, Multi-Year Ambassadorial Scholarships, and a Cultural Ambassadorial Scholarship for language study. There are also Rotary grants available for teachers. Consult your local Rotary club to find out which ones you qualify for.

- Academic-Year Ambassadorial Scholarship
  The Ambassadorial Scholarship awards up to $25,000 USD to study abroad for an academic year. The most common type of scholarship offered, last year Rotary awarded one thousand of these scholarships.

- Multi-Year Ambassadorial Scholarship
  The Multi-Year Ambassadorial Scholarship awards $12,000 USD per year for up to two to three years of degree study overseas. Last year, 150 scholarships were awarded, mostly by Rotary districts in Japan and Korea.

- Cultural Ambassadorial Scholarship
  This program awards up to $12,000 USD for three months, or $19,000 USD for six months, for intensive language and cultural training, including homestays. On a Cultural Ambassadorial Scholarship you can study Arabic, English, French, German, Hebrew, Italian, Japanese, Korean, Mandarin Chinese, Polish, Portuguese, Russian, Spanish, Swahili, or Swedish.

- Rotary Grants for University Teachers
  These grants are available to higher education teachers willing to teach at colleges and universities in developing countries. You can apply for a three- to five-month teaching grant of $12,500 USD or a six- to ten-month grant for $22,500 USD. Forty-seven grants were awarded in 1999–2000 and thirty-six grants were awarded in 2001–02.

FEDERAL GOVERNMENT SCHOLARSHIPS AND LOANS

The biggest government-supported federal scholarship program is the Fulbright Program sponsored by the Institute of International Education (IIE). Fulbright scholarships are available to college students, graduates, postdoctorate scholars, and professionals from a wide variety of fields.

Students who receive financial aid from the government may be able to use it toward a study-abroad program. If you are studying abroad on a program offered by a university other than your home university and you want to apply your government aid, you must complete a consortial agreement that states you will receive aid money from either your university or the sponsoring university, but not both.

★ **INSTITUTE OF INTERNATIONAL EDUCATION (IIE)**

www.iie.org

3007 Tilden Street NW, Suite 5L
Washington, DC 20008
Phone: 202-686-4000
Fax: 202-362-3442

The Institute of International Education is the world's most experienced educational exchange and training organization, offering a host of programs for students and professionals, including many scholarship opportunities. The Fulbright Program, which is a program of IIE, is the main scholarship program sponsored by the U.S. government. The U.S. Department of State is the principal administrator.

★ **COUNCIL FOR INTERNATIONAL EXCHANGE OF SCHOLARS (FOR POSTDOCTORATE SCHOLARS)**

www.cies.org

THE FULBRIGHT PROGRAM (FOR POSTBACCALAUREATE DEGREES)

www.iie.org/cies

THE FULBRIGHT PROGRAM (FOR UNDERGRADUATES)

www.iie.org/fulbright

Contrary to some of the myths you might have heard about the Fulbright Program, scholarships are available to scholars as well as students, lawyers, journalists, musicians, filmmakers, scientists, engineers, and artists to study, teach, lecture, and do research in one of 140 countries. The program was designed to promote self-development through international experience, so participants may create their own research agendas or special projects. For example, you could study contemporary artistic expression in India, research women's rights in Chile, or do cancer research in the U.K. Some programs require a foreign language, but others do not.

Programs range from two months up to one year. Full grants are offered and include transportation, living and research expenses, and health and accident insurance. Travel grants to Germany, Hungry, and Italy are offered to supplement a non-IIE sponsored scholarship award. Foreign and private grants are also available. Alternatively, you could apply for a grant to teach English or a business grant to research and intern abroad. You can search grant summaries on the Fulbright Web site by country. Eligibility requirements vary per program, but are generally not as strict as other scholarship programs.

IIE has many other scholarship opportunities besides the Fulbright Program. Here are some of their additional offerings.

★ Emerging Markets Developers Advisors Program (EMDAP)

www.iie.org/pgms/emdap

EMDAP provides American graduate business students with the opportunity to help manage small and medium-size enterprises in developing countries for a period of ten months. Applicants must be U.S. citizens currently enrolled in an accredited graduate program and must complete business-degree course work upon returning from abroad. Preference is given to candidates who have two to three years of management experience before entering a graduate business-school program, academic achievement, foreign-language competency, a commitment to the developing world, and the maturity to work unsupervised. The program covers travel, living, housing, and insurance expenses.

★ Freeman-Asia Scholarships

www.iie.org/pgms/freeman

To be eligible to apply for this scholarship, you must be a U.S. citizen studying at the undergraduate level at an American accredited institution and you must demonstrate financial need to study in Asia. Scholarships are awarded in the amounts of $3,000 USD for a summer program, $5,000 USD for a semester, and $7,000 USD for an academic year.

★ Investing Women in Development (IWID) Fellows Program

www.iie.org/pgms/iwid/overview

The application process for this program is open to U.S. citizens with an M.A. degree in a relevant target area and four to fifteen years of work experience in one of the following areas: democracy and governance, business and economic development, girl's education, environment, or health and nutrition. Qualified applicants will have the opportunity to obtain hands-on development experience in an international setting for up to twelve months. Fellows receive a personal stipend, a housing allowance, insurance, and airfare.

★ Lucent Global Scholars Program

www.iie.org/pgms/lucent

To apply for this program, you must be enrolled in your last year of secondary school and have distinguished yourself in math and science. Students outside the United States must be enrolled at a participating university, be in their first year of studies, and no older than twenty-one. Second-year university students may be eligible in some countries. Winners of the Global Scholars Science Competition receive a one-time financial award of $5,000 USD.

★ Midwest Educational Associates Scholarship

www.iie.org/midwest/grant

Competition for this scholarship is open to undergraduate students currently enrolled at a Midwestern college or university that is a member of

IIE's Educational Associates Program. One $4,000 USD, three $2,000 USD, and several $1,000 USD study-abroad scholarships are available.

National Security Education Program (NSEP) Boren Scholarship
www.iie/org/nsep

To be eligible for a Boren scholarship, you must be a U.S. citizen studying at the undergraduate level at an American accredited institution, and planning to use the scholarship money to study abroad. Awards are based on economic need. The minimum award amounts are $2,500 USD for a summer program, $4,000 USD for a semester, and $6,000 USD for a year. The maximum award for one term is $10,000 USD and $20,000 USD for one academic year.

ADDITIONAL SCHOLARSHIP RESOURCES

To find out what other types of scholarships and funding means are available for your field of study abroad, search the following resources:

FastWeb
www.fastweb.com

Search for scholarships (and for your dream college) for free.

International Education
http://internationaled.about.com/education/internationaled/cs/financialaid/index.htm

A link to over seven hundred sites dealing with adult education, distance learning, financial aid, language study, study abroad, and more.

International Student
www.internationalstudent.com/scholarship_search.htm

Enter your field of study and get a list of available scholarships.

The Scholarship Experts
www.scholarshipexperts.com

A scholarship search for U.S. and international students. There are fees for the service, which cost $29.95 USD or $49.95 USD, depending on which plan you choose. The silver membership ($29.95) gives you unlimited search results for a full year in the national and state award categories. The gold membership ($49.95) gives you unlimited search results for a full year in the national, state, local, and institutional award categories.

★ **FINANCIAL RESOURCES FOR INTERNATIONAL STUDY: A GUIDE FOR U.S. STUDENTS AND PROFESSIONALS**
(Institute for International Education, 1996)

Lists seven hundred fellowships, grants, scholarships, and paid internships for undergraduate to postdoctoral students and working professionals.

Fund-Raising

Fund-raising is a creative pursuit requiring out-of-the-box thinking. You can raise funds in countless ways; however, fund-raising basically boils down to either asking people for contributions or finding ways to earn money yourself. Most organizations won't simply give money away, but they will usually listen to creative ideas for mutual benefit. For example, your local sporting goods store might be willing to pitch in a few hundred dollars and some clothing for your travel adventures if you offer to clean up the storeroom and work an inventory shift. Some ideas, like garage sales, are so obvious many of us overlook them as potential fund-raising schemes.

Here are some ideas to get you into a fund-raising state of mind.

ASKING FOR MONEY

First you have to figure out *who* to ask. Possible contributors include local, regional, and national organizations, as well as family, friends, and colleagues. Check each potential organization's Web site and find out if you might already qualify for scholarships, grants, or other funding. Here are a few organizations you can begin to target:

- Campus groups
  There are many associations on campus that have funds to help both members and nonmembers, including academic clubs, sports teams, campus groups, and local and national chapters of sororities and fraternities.

- Civic groups
  Most towns, whether large or small, have local branches of Rotary, Kiwanis, Elks, Lions, or Quota clubs. If you or someone you know belongs to one of these groups, you may have a better chance of getting funding for your overseas travels.

- Local companies
  Your first stop should be your own company (or your husband's or sister's or dad's)—do they offer a scholarship or grant that you qualify for? If not, try companies in your area. They usually like to support their community's citizens, especially if that support generates some publicity. Or find a company with an international division. If you are going abroad to study international business, promise to volunteer a certain number of hours in their international department in return for a study-abroad donation. Better yet, if the company has an office in your targeted country or city, maybe they'll sponsor you in exchange for your services over there.

- Heritage groups
  Heritage groups like the Italian American Heritage Foundation (www.iahfsj.org), the Supreme Lodge of the Danish Sisterhood of America (www.danishsisterhood.org), and the National Japanese American Historical Society (www.nikkeiheritage.org) often offer scholarships, grants, or funds to further the cultural and linguistic education of their members. Check with your own ethnic organization to see if they offer similar opportunities.

- Alumni associations
  Most high schools and colleges have alumni associations. Although they usually don't have a great deal of money to spare, it never hurts to ask.

- Media organizations
  Newspapers and radio and television stations are good potential funding sources, especially if you convince them that your travel adventures will make a good "local color" story upon your return. Imagine, you might even gain a bit of fame!

- Sister cities
  If your city or town has a sister city in a country you would like to visit, you may be able to get some funding dollars from either your city or the sister city in honor of furthering the cultural exchange between the two.

- Friends, family, neighbors, and colleagues
  When you tell your family, friends, and coworkers about your fund-raising efforts, many of them will gladly pitch in a few bucks to support you. However, if they don't offer to contribute and you feel strange asking them for money outright, try approaching it in a different way. Ask for financial contributions instead of presents for birthdays and the holidays. Or promise that anyone who contributes to your fund-raising efforts will be treated to a homemade ethnic dinner (complete with take-home recipes) upon your return.

And the ideas don't stop there. Foundations, professional and business clubs, charitable organizations, religious groups, travel agents, and almost any organization listed in the phone book might be able to contribute financially for your trip. It all adds up, no matter how big or small the contribution.

Once you've narrowed down your list of potential funding partners, write an enthusiastic letter explaining your intentions to study, intern, volunteer, or work abroad.

- Be Direct, Personal, and Courteous
  Communicate your long-range career or life aspirations in a way that reflects how serious and determined you are. (No one wants to fund someone who might

not make good use of the funding.) Make a connection between the organization's mission and your goals, and explain how you can help each other.

- Explain Clearly How the Funds Will Be Used
  If you are trying to raise $1,500 USD to pay for your flight to volunteer in Rwanda, or $10,000 USD to pay for tuition to study abroad in Japan, be sure to make your goals known. Briefly explain your fund-raising strategy, including your target amount and deadline date.

- Offer Something Valuable to the Organization
  In return for the funding you are asking for, offer to send regular updates of your experience overseas, highlighting what you are learning and how the experience is impacting you. This way the contributor can see how their funds are making it possible for you to achieve your goals. Or arrange a slide show or video presentation upon your return to educate your sponsors and the public about the culture and the political and economic issues of the country you visited. A slide or video show could also be part of a seminar or discussion you lead at an on-campus event, or wrangle your way into a radio or television interview. See chapter twelve for more ways to share your experiences with your funding sources or community while you are overseas and when you return home.

- Include a Photo of Yourself in Your Letter
  Giving the organization a face to put with the name will help you establish a personal connection with them. The organization will more seriously consider your request if you are personal and genuine. If they think you just want their money, they'll likely reject your request for funding.

- Stay Persistent, Determined, and Positive
  You will receive rejections, but don't take them personally. Some organizations that would like to help simply may not have the extra funds to do so. And when you are successful, be sure to express appreciation and support for the organization's generous contribution.

RAISING MONEY

When the money you earn through sponsorships and contributions isn't enough to cover all of your costs, you might have to earn the rest with the help of some friends. One or two ideas is all you need to get started. Besides the obvious solution of getting a part-time job, here's what has worked for other ambitious fund-raisers.

- Have a Bake Sale, Garage Sale, or Rummage Sale
  Bake sales, garage sales, and rummage sales are relatively easy and fun ways to earn more than just a few dollars. Sell your favorite cookies or get rid of all those things in the garage that you don't need. Also ask friends if they have food or household items they would like to contribute to your cause. Pick a nice day on the weekend, hang a few signs around the neighborhood, and place an ad in the local newspaper. Provide educational information about the country you'll be visiting and tell people the sale is a fund-raising event. Gladly accept donations.

- Sponsor a Car Wash, Do Yard Work, or Paint Houses
  At the very least family members and friends will bring their cars around for a good washing. And if you advertise your car wash, the profits are sure to increase. Likewise, your neighbors might enjoy a respite from mowing the lawn, trimming the bushes, or other yard work. And if you're really ambitious, offer to paint their house. If you do a good job, their contribution to your cause might be quite generous.

- Give a Class or Seminar
  The local YMCA or your health club could be your collaborators for this project. Find out if you can sponsor a series of special classes, like yoga, dance, or basketball to raise money for your trip. Your sports center would benefit from the additional course offerings and exposure, and you would meet new people, keep in shape, and get that much closer to your financial goal.

- Collaborate with Your Church or Religious Group
  Religious groups tend to hold their own fund-raising and community events, like

dinners, raffles, and bingo nights. Ask group leaders if you can give a talk at one of these events (perhaps about the country you'll be visiting) to raise donations.

- Sell Your Used Computers, Software, or CDs
  Collect used computers, software, CDs, books, and other technological or educational materials from friends, neighbors, and family members. When you explain that it's for a good cause, chances are you'll accumulate a carload of goods pretty quickly. Then ask local high schools, community colleges, and other educational institutions if they would like to buy some used computers, software, and books for a great price. You'll unload your car in no time—and make a pretty good profit.

Besides asking for contributions and earning the money yourself, how else can you get what you need for your trip without spending a dime? Just ask.

ASKING FOR WHAT YOU NEED

It is so obvious but some people never consider going directly to the source to ask for what they need. The worst that can happen is someone says "no," but it never hurts to ask. You might come away with a new travel wardrobe, a gift for your hosts, or a discounted plane ticket. As a courtesy, it is a nice gesture to at least send a postcard to the people, stores, and organizations that donated items for your trip. Or drop in for a visit when you return home.

- Stores
  Ask clothing and sporting goods stores if they would like to donate backpacks or suitcases, clothes, or camping gear for your excursion overseas. Many stores have overstock or sale items that they will donate without impacting their bottom line.

- Airlines
  Ask airlines for a discounted or free ticket to your destination. You need to be clever and package your request as a win-win opportunity for both of you. Offer to help with their promotional events or write an article about your destination for their inflight magazine. Once again, it never hurts to ask.

- Banks and Family Members
  They've got it. You need it. Money! Banks are there to loan you money and often have good interest rates and terms for students. You might also convince family members to loan you the money if you agree to pay it back within a certain time period upon returning from your trip.

- Gift Shops
  It is always a good idea to bring a few gifts from home to share with your hosts or people you meet during your travels. Your local gift shop might be willing to donate gifts that are particularly representative of your community, state, or country, such as postcards, picture books, and handmade crafts.

- Camera Stores
  During your overseas travels, you'll want to capture the local culture on film or video. That means you'll need a camcorder, camera, batteries, film, photo albums, and other accessories. Your local camera or photo store might be willing to chip in a few items, at least a few rolls of film. In return, you might offer some of your travel photos for their displays.

ADDITIONAL FUND-RAISING RESOURCES

★ **THE AMERICAN INSTITUTE FOR FOREIGN STUDY (AIFS)**
www.acis.com/students/fund-raising.asp

Toll-free: 800-888-2247
Phone: 877-474-9343

AIFS has a long list of fund-raising ideas for students, from bake sales and car washes to breakfast in bed.

★ **MICHIGAN STATE UNIVERSITY (MSU)**
http://study-abroad.msu.edu/shared/afford.html

MSU has a very good study-abroad Web site with valuable financing tips.

One last bit of advice: start fund-raising early. The sooner you begin, the more money you'll raise.

A GLOBAL CITIZEN'S BEST BETS FOR TRAVEL DISCOUNTS

American Association of Retired Persons (AARP)

www.aarp.org

Toll-free: 800-515-2299

AARP offers travel specials and provides important travel information to the fifty-plus crowd. For example, as an AARP member you can travel at a special AARP Savers rate on U.S. Airways. You can also get answers to frequently asked questions about moving overseas on their Web site.

The American Institute for Foreign Study (AIFS)

www.acis.com/trips/adult.asp

Tel: 800-888-2247 or 877-474-9343

The American Institute for Foreign Study (AIFS) offers adult travelers free or discounted travel by becoming group leaders. When you sign up a group of six people (friends, relatives, or strangers) for any of their select tours, you will receive a free trip or a generous cash stipend, along with other benefits. (First-time group leaders only need five participants.) What a deal!

Council Travel

www.ciee.org

Toll-free: 800-2COUNCIL

Council helps students arrange for student identity cards, which make them eligible for discounts worldwide, including a whopping 50 percent discount on airfares.

TEN EASY WAYS TO RACK UP FREQUENT-FLIER MILES

1. Sign up to be a member of the frequent-flier program on every airline you fly. Even if you don't have enough mileage for a free ticket, you may be able to put part of your miles toward the total ticket price, thus lowering the price you pay. If you stay loyal to one or two particular airlines, you may qualify for their exclusive platinum or gold clubs, making you eligible for even more benefits.

2. Rack up the miles by booking your ticket through the airline's Web site. By booking online, you may earn an additional 1,000 miles or more. Watch for special offers for booking to certain destinations, staying in affiliate hotels, or renting cars from one of the airline's rental car partners.

3. If you don't fly frequently or don't currently have miles earned with any airline, you can start racking up miles by getting a credit card that gives you miles for every dollar you charge on your card. I use the Citibank Visa AAdvantage card, which immediately applies my miles to American Airlines. But there are credit cards affiliated with most of the airlines these days. Membership usually offers travel discounts on car rentals, restaurants, and hotel packages too. You can choose a combination airline/hotel card, a charge card, a bank credit card with travel advantages, or a debit card. To find the best card for you, see www.frequentflier.com/card-intro.htm.

4. Financial institutions have also started offering their customers frequent-flier miles. For example, you may be able to pay for your mortgage with a credit card that earns miles, or earn miles directly by investing in dedicated investment vehicles, like the American AAdvantage money market fund.

5. Phone companies are partnering up with airlines too. Use a long-distance phone company that is linked with the frequent-flier program of your choice. I use MCI Worldcom, which gives me five miles for every dollar I spend with them. Of course, I pay MCI with my American Airlines credit

card to get even more miles! Alternatively, you may be able to buy a phone card, such as the American Airlines frequent-flier phone card, which gives you five hundred free miles when you purchase the $45 USD phone card.

6. Buy online with companies that are connected to frequent-flier programs. Check out www.Clickmiles.com to order books, clothes, and pharmacy articles through their online retail partners. You'll get miles for every dollar you spend (and they often have very good offers). And again, paying for items with your credit card will give you at least double the miles.

7. Similar to frequent-flier programs, always sign up with hotel and car rental companies' points programs. They could get you discounts, free hotel stays, free car rentals, and other perks.

8. Track your miles through www.maxmiles.com. Every time you show your frequent-flier card to the airline you are flying with, the miles will show up automatically on your Max Miles account. Members can also take advantage of various travel promotions.

9. Watch the Internet for bargains. Pay attention to airline auctions, special offers, and price wars to get cheap tickets and increased frequent-flier miles. You can also try www.hotwire.com or www.priceline.com to name your ticket price. Remember to use your frequent-flier credit card to pay for your ticket purchase.

10. To manage your frequent-flier miles, find deals to earn more miles, and track worldwide flight schedules, use one of the top frequent-flier Web sites, WebFlyer, at www.webflyer.com.

Airlines continue to form new partnerships with a variety of companies to make it increasingly easier for you to travel. Be sure to keep abreast of your favorite airlines' special frequent-flier offers.

CHAPTER TEN

So You Want to Go Abroad? Predeparture To-Dos

The reward of a thing well done is to have done it.

RALPH WALDO EMERSON

Some of the most common questions people ask about going abroad concern logistics: What do I do first? How do I communicate if I don't know the language? How will I meet people or get a job? Going abroad doesn't have to be difficult. In fact, with the right attitude and a little pretrip preparation, it can be as fun and exciting as you imagine it will be. This chapter will help you get the practical things done so your global adventures are as easy and painless as possible.

Predeparture Preparation

Gearing up for your trip overseas involves a bit of advance planning. Working through the following list will make this step easy.

GET A PASSPORT AND A VISA

Before you can even step on the plane to your global destination, you'll need a passport. Regardless of your nationality, a passport is required to travel to most foreign countries. There are over forty-five hundred passport acceptance centers and thirteen passport agencies in the United States, as well as post offices, libraries, county and municipal offices, and federal, state, and probate courts where you can apply for a passport. Apply several months before your departure (earlier if you also need a visa from a foreign embassy) as it generally takes five to six weeks for delivery. A visa is a legal stamp or document that you buy from a foreign government to be able to enter that country. Some countries bar people of certain nationalities from entering their countries without a visa. Visas can be obtained from the embassy of the country you are visiting. In some cases, you may be able to buy your visa at the country's airport upon arrival. That country's embassy or a travel agent can provide you with details on the procedure and cost. For more information about passports and to find the closest passport agency or acceptance center near you, visit http://travel.state.gov/passport_services.html.

FILL UP YOUR BANK ACCOUNTS

Before you leave home to travel or live abroad, make sure you have enough money in your bank account to cover your initial expenses. Be sure your bank's ATM card is accepted at ATMs worldwide so you can withdraw cash from your checking or savings accounts. As most international ATMs accept only four-digit PINs, you may have to consult with your bank to make sure your card will work properly overseas. Also, if you will be transferring money to your bank account while abroad, get your bank's routing information and SWIFT number for money transfers. (SWIFT is a special routing code for international bank money transfers.)

BUY TRAVELER'S CHECKS

Traveler's checks are designed to be used instead of cash. If you lose your money while traveling, you're out of luck. But if you lose a traveler's check, you can get your money back, as long as you have your receipt. Traveler's checks are also good things to have if you happen to lose your wallet (as long as the checks aren't in your wallet), if you don't want to use a credit card or cash to pay for things, or if you simply want to have a financial backup when traveling. There are usually small fees (1 to 2 percent) charged for buying traveler's checks, however AAA members can get them free of charge at any AAA office, and some banks will waive the fees if you have a minimum balance or a particular type of account with them. And if you buy American Express traveler's checks, you can cash them for local currency at American Express offices worldwide without incurring any exchange fees.

CHECK YOUR CREDIT CARDS

If you don't already have a credit card, get one, or even a few. Visa and MasterCard are accepted more widely than American Express, but whatever credit card you decide to get, make sure it is linked up with a frequent-flier, hotel, and car-rental program. For your protection, get a credit card that will include your photo on your card to prevent anyone else from using it. And make sure you know how to replace your card if it gets stolen. The easiest way to do this is to make a copy of the front and back of your card and keep the copy in a safe place, separate from your card.

CONSIDER A DEBIT CARD

I have a direct debit Visa card that allows me to use almost any ATM in most European countries to withdraw cash from my money market account in the States. My money earns more interest in a money market account than it would in a normal savings or checking account. Check with your bank or financial institution to see if they offer debit cards, and find out where the cards are accepted worldwide.

ORGANIZE YOUR BILLS

If you are moving abroad for an extended stay, you'll probably want to cancel your services. When you call the phone and utility companies to cancel, arrange to have the last bill sent to the address of a friend or family member who can pay the bill for you. Leave enough money in your bank account to cover the bills and sign a few checks to leave in the hands of someone you trust. If you do not plan to cancel your services, another alternative is to have the amount of your bills automatically withdrawn from your bank account. Citibank Online (www.citibank.com) has an automatic bill paying service that withdraws money from your bank account and pays your bills so you do not have to write checks. This is also a good option for recurring charges such as mortgage or car payments that you may still have to pay while living abroad.

About one to two weeks before you go abroad, discontinue the rest of the services you won't need while you are away. These services could include newspapers and magazines, mail delivery, housecleaning, and anything else you can think of.

COLLECT IMPORTANT ADDRESSES AND PHONE NUMBERS

You've already gathered the addresses of friends and relatives you're going to send a postcard to, but have you remembered to include other important numbers you might need, like your doctor, your insurance company, your bank, your broker, your property manager, your secretary, or anyone else with whom you do business? Depending on your reason for traveling overseas, you might also include contact information of the embassy and consulate of the countries you'll be visiting, the hotel or place you'll be staying, and the company or organization where you will be working. Don't forget to leave all of this information with Mom, Pop, and your best friend in case someone at home needs to contact you in an emergency.

GET AN INTERNATIONAL CALLING CARD

An international phone card can save you money when calling home or between foreign countries. With an international calling card from AT&T, MCI, or one of the many other carriers, you can call home by punching in an access code for the country you are calling from. If you are part of an international calling program, your phone rates for calling overseas will be discounted. Calling the States from hotels and private phones abroad can be very expensive, and at the time of this writing, most U.S. mobile phones do not yet have international coverage. However, as technology develops and international calling rates drop, you should compare calling-card rates.

INVESTIGATE INSURANCE OPTIONS

It's always better to be safe than sorry when traveling abroad. Should the worst happen—you lose your luggage in India, suffer a bad stomach sickness in Ghana, or get caught in a political conflict while working in the Middle East—you will feel more secure when you know that your situation is covered by insurance.

When you go abroad to travel or live for a long period of time, make sure your insurance covers the basics. Insurance offerings vary, but you may want to buy travel insurance that covers trip cancellations, travel delays, emergency evacuations, baggage loss or delay, rental car protection, and basic health problems. Depending on the political stability of the region you are visiting, you might also consider emergency evacuation insurance for you and your family in case of an uprising, coup, or other potentially threatening situation.

Your health insurer or company's benefits department should be able to offer you plans that will include comprehensive medical benefits, worldwide access to the best health-care facilities, and evacuation in a health emergency. Student travelers can purchase special travel and health insurance packages through travel agents, insurance providers, and even exchange programs like Council Travel, a division of CIEE/Council on International Educational Exchange (www.counciltravel.com). Nonstudents may have to look into individual coverage offered by insurance companies. Coverage can be expensive, but short-term plans are available that may help you save on costs. If your company is sending you overseas, they should be responsible for arranging international travel and health insurance for you and your family.

SQUARE AWAY YOUR INVESTMENTS

Most banks and financial institutions are making it increasingly easy to take care of your banking and investing needs online. Unless you want to deal with calling and faxing your bank from overseas, I suggest learning how to invest online. My own accounts are handled by Charles Schwab at www.schwab.com; however, there are a number of online banking and investing companies available.

PREPARE YOUR FAMILY

Traveling with a family (especially young kids) can be a challenge and might take some extra planning. However, my husband and I traveled with our four-month-old son and it wasn't nearly as bad as we thought it would be. If you have a small baby, request a bassinet and ask to sit in the bulkhead seats. If you have older kids, they should have their own seats and plenty to keep them busy. The airline may provide a coloring book and crayons or toys to play with, but you might also want to supplement those things with snacks and your kids' favorite diversions.

PACK UP YOUR PETS

Can Fido go with you on your trip? Most countries have strict rules regarding pets, especially island countries like England and Australia, which claim to be rabies free. In most cases you'll have to subject your pet to a period of quarantine, which could end up being longer than the amount of time you're planning to be abroad. All of your furry friend's shots must be up to date and you must get an official certificate signed by your veterinarian within a particular time period prior to your trip.

If you're traveling to a very hot or very cold climate, you might also have trouble booking your pet's passage on a plane. Many airlines have temperature restrictions for pets because the cargo hold isn't temperature controlled. To find out what the pet regulations are for a particular country, contact that country's embassy or consulate.

MOVE OR STORE YOUR BELONGINGS

Going overseas for a long period of time to study or work might entail moving or storing the stuff you can't fit into your suitcases, like boxes of books or furniture. If you are being transferred overseas by your company, they will likely arrange for movers and pay the cost. But if you have to foot the bill on your own, compare the costs of sending your goods by land, by sea, and by air through Federal Express, DHL, or a moving company like Graebel. To ensure your moving company is nationally approved, choose a company that is a member of the internationally recognized FIDI moving association at www.fidi.com.

If you're leaving your possessions behind, you'll have to find a safe place to store them. The cheapest way is to ask a friend or relative if they have some extra storage room in a garage or attic. If not, you may have to pay a storage company to store your things. Depending on how attached you are to your belongings, you should weigh the costs of storing your things for the amount of time you'll be away with the benefits of selling your stuff and buying new things when you return.

FILE YOUR TAXES

If you are an American citizen working overseas, you will still have to file taxes, but you may be able to get around paying much, if anything, to Uncle Sam. The Foreign Earned Income Exclusion is an annual tax allowance (under section 911 of the Internal Revenue Code) that allows Americans working abroad to exclude up to $80,000 USD for fiscal year 2002. (Check with the IRS for exclusions for subsequent years.) You must have lived abroad for at least 330 days, and you will need to file Form 2555 with your 1040 form to get the exclusion. To read more, log on to www.irs.ustreas.gov/prod/forms_pubs/pubs/p54ch04.htm, or order Publication 54 from the IRS. Consult a tax advisor to find out how much money you can save by taking advantage of this exclusion. And even if you aren't eligible for this exclusion, make sure you know how to deal with your taxes overseas once tax season rolls around.

Other Preparations

Going abroad takes an investment of valuable time, energy, and sometimes money. Whether you go abroad for academic, professional, or personal reasons, knowing how to make the most of your investment will help you maximize your opportunities abroad and apply what you learned upon your return (if you return).

Your company has just offered you an exciting opportunity to work in their São Paulo office. It's the opportunity of a lifetime and you can't wait to go. You might be tempted to throw your clothes in a suitcase and hop the next plane to Brazil. Many people have done just that and later wished they had better prepared for their adventure. Here are some ways to get ready for your life overseas.

TEST YOUR CULTURAL COMPATIBILITY

Before you set expectations for your trip, do a cross-cultural compatibility test (see page 34) to see how your personality and values compare with those of the country you'll be visiting. This is where the results from your personality and cultural assessments come in handy. It is not always necessary to choose a country whose values align with yours. It can be just as interesting to discover a land and people that are different than what you are used to.

However, it is better to find out (ahead of time) acceptable ways of interacting in society and business than to find out with a cultural clash. If you are used to working in a goal-oriented, orderly fashion, you might be in danger of clashing with your coworkers in Argentina, Brazil, or Turkey, who have different approaches to work. Richard Lewis Communications (www.crossculture.com), experts in the cross-cultural training field, offer several software products for individuals and businesses that analyze cross-cultural compatibility, offer effective ways of handling potential areas of conflict in business and social situations, and allow you to create your own cultural profile for future use. A book I found particularly helpful is *When Cultures Collide* by Richard Lewis (Nicholas Brealey Publishing, 1999).

LEARN ABOUT THE COUNTRY AND REGION

Buy a travel guide. For long stays abroad, I recommend *Eyewitness* travel guides. For short stays, where you might not get a chance to leave the city or country in which you are living, a city or country guide should be sufficient to help you plan your activities. Your travel guide should include interesting facts about the country, cultural information, a calendar of holidays and cultural events, sight-seeing information, travel itineraries, and other tidbits that help you adjust to your new home quickly. As you read about your new country, draw up a plan of all the museums, cultural sites, traditional festivals, cities, outdoor activities, and other things you definitely want to see and do during your stay. Whether your stay will last three months, two years, or indefinitely, it is always good to know about what is going on in the area so you can make the most of your time there.

START LEARNING THE LANGUAGE

Even if you find out you will be leaving for Japan next Thursday, you still have plenty of time to learn at least five Japanese sayings. Invest in language CDs, cassettes, or software. Language learning is particularly fun and easy with interactive software (and you can even play it on your laptop during the plane ride). Good language learning tapes and software take you beyond the basics, and they include business vocabulary too. Berlitz (www.berlitz.com) has an assortment of books, tapes, software, and CDs in many languages at most levels. If you have more time to prepare for your trip, take a language class with a native speaker. Classes generally teach important cultural information that you may not get from a language CD or software, and you can ask your teacher lots of questions. Learning as much of the language as you can before you leave home will give you a jump start on acclimating yourself upon arrival at your destination.

NETWORK WITH OTHER GLOBAL CITIZENS

To launch you into the international life you've always dreamed about, talk to people who feel the same way you do. People who have never wanted to go where you are going simply won't get the "international thing." Start by hooking up with like-minded international globetrotters. Seek out people who live or have lived in the country you'll soon be calling home. Ask your friends, colleagues, family, and language teacher for recommendations. Search your local community, school or workplace, and online for people with international experience. Once you meet people who can help you, pick their brains about the cultural dos and don'ts, things they liked and didn't like, travel tips, and even restaurants they recommend. Write this insider info in a journal so you can refer back to it later. By the way, make sure you add these people to your international list of networking contacts, and keep in touch with them.

FIND A MENTOR

If you have very specific global aspirations, seek out the advice and inspiration of someone in your field who has successfully done what you plan to do. If this person is someone you really admire, take this relationship one step further and ask this person to be your international mentor. As such they will help guide you to your global destination, whether that be a corporate career with a multinational company or a personal mission to help a world cause. Also seek out the advice of international career counselors who specialize in your field. Local colleges and universities should be able to recommend international career counselors in your area, or you can always check the yellow pages.

MANAGE YOUR EXPECTATIONS

Before you depart, write down your expectations, hopes, and fears and discuss them with anyone who will be sharing the experience with you, like your spouse and children. Be aware of the challenges you and your family will be facing—climate, mentality, religious differences, conveniences like supermarkets and quality medical care, standards of living, access to social and work opportunities—and acknowledge that you will all change as a result of your experiences. Culture shock is usually felt most intensely during your first extended stay abroad, but it can happen during any stay abroad, depending on how well you are prepared for the environment. Feeling a wide range of emotions, from intense excitement to deep frustration and disappointment, are normal until you learn how to reach a happy balance.

Once you feel you've achieved a perspective on your life overseas, you may return home to feel your life turn upside down again. As your international assignment winds down, you may be looking forward to returning to your old life, only to find out upon your return that your old lifestyle doesn't satisfy you like it once did. Reverse culture shock may leave you feeling out of sync with your native environment and friends. It will take time to put your international experiences into perspective in your homeland, but the more you prepare for it now, the easier it will be when you return. Also, realize that you now belong to a community of individuals who, during the course of their own global adventures, have dealt with similar emotions and have enjoyed the rewards of a global life.

THINK ABOUT REPATRIATION

If you go abroad with your company, ask about repatriation assistance before you leave for your overseas position. Repatriation is the process of transitioning back to your home country and company after living abroad for an extended period of time. Some companies deal with repatriation issues, but many don't. In fact, many companies that send employees abroad fail miserably with their repatriation attempts.

Most of the time employees must rely on their own resources to adequately prepare themselves for repatriation. Your company may not guarantee you a challenging position upon your return and you may find yourself looking for a new job in a country you haven't worked in for several years, with a job market that's dramatically different than when you left. Keeping in touch with the professional job front and contacts in your home country while you are overseas will put you one step ahead when you decide to come home again. Before and during your overseas assignment, think about where you want to live when you return home, what kind of work you want to be doing, and whether your company will have a position for you or if you will have to embark on a new job search.

Planning at least a few months before your departure date will help you avoid last-minute, stressful situations. In fact, as you start working through the predeparture checklist, your excitement about going abroad will build and all of the final details will gradually fall into place. Before you know it, you'll be saying your last good-byes and hopping on the plane!

CHAPTER ELEVEN

You're There! Now What? Maximizing Your Time Abroad

Do not go where the path may lead,
go instead where there is no path
and leave a trail.

RALPH WALDO EMERSON

You've done all the research, you completed the predeparture checklist, and you've just landed in your new overseas home. Now what?

First Things First

Whether you're going solo or whether your company or an exchange program has all the logistics planned for you, there are some basic things you'll need to take care of once you step out of the plane.

EXCHANGING MONEY

If you brought cash or traveler's checks, you'll probably want to find an exchange booth, American Express office, or bank to exchange your U.S. dollars for the local currency. Always compare exchange and commission rates, as they vary significantly. Some banks try to rip you off by charging an unheard-of 20 percent commission to exchange cash sums, while American Express charges nothing to exchange traveler's checks. Most Western countries have ATM machines that charge about the same for withdrawals as your bank at home. International airports always have exchange offices, but not always the best rates. Nevertheless, I recommend getting at least a small amount of local currency before you leave the airport.

SECURING GROUND TRANSPORTATION

Off the plane, out of the airport, and now you need to get somewhere. If no one is picking you up at the airport and you are not renting a car, you'll be at the mercy of local transportation—taxis, buses, trains, rickshaws, camels, whatever! Chances are you'll be taking a taxi to your hotel or wherever you'll be spending your first few nights. Try to avoid getting ripped off by taxi drivers: they use every trick in the book. Read ahead in your tour book for tips about the local transportation. If you are traveling to Eastern Europe and Russia, for example, know how to recognize and avoid the "mafia" taxis. In general, taxis in any country should have prices posted in the window that are consistent with taxis from other companies. They should also have a phone number plastered somewhere on the outside of the taxi. Just make sure the meter is working properly and the driver isn't taking you on the scenic route, unless you want him too. If you have a good map of the city, double-check his route.

If you have to catch a train upon arrival, many international airports have direct train connections. Otherwise, you might have to take a taxi to the main train station. If you haven't purchased your rail ticket in advance, you'll have to locate the ticket counter for trains to your destination. There could be different ticket counters for local trains and intercity trains. Seats on busy routes may sell out quickly, so consider paying the extra fee to reserve a seat. You might also consider traveling first class, especially if the second-class seats get crowded or train safety is questionable in the country in which you are traveling. Students and seniors might be eligible for discounts. If you need to catch some shut-eye on the train, especially after a long flight, reserve a seat in a sleeper wagon, where the seats pull down into beds. Most of the time it is safe to sleep on trains, but use caution and common sense. Falling asleep on the train can leave you and your belongings vulnerable to thieves. Don't forget to safely tuck away your wallet, passport, traveler's checks, cash, and any other valuables inside your clothes and close to your body.

The bus is another way to get to where you are going. From the airport, you may be able to get bus service to city destinations, and sometimes to other cities within the country. Otherwise, you may have to take a taxi or train to the main bus terminal. Bus tickets can usually be purchased ahead of time at a ticket booth or ticket machine. If you will be using local bus service for several weeks or months, consider buying a special week or month pass to save money. Discounted tickets might be available for students and seniors. Bus safety varies per country, but it is always better to err on the side of caution. Buses, especially crowded ones, are a great place for pickpockets. Before you get on the bus, make sure all of your valuables, including money and passport, are safely tucked away on your person so that a thief cannot reach them. Hold on to your belongings at all times.

FINDING HOUSING

If you are not going abroad through an exchange program that arranges housing, you've got to get a roof over your head right away. For temporary housing solutions, stay with a friend or at a youth hostel, pension, or hotel. It is a good idea for independent travelers to have their quarters booked ahead of time. To find vacant apartment rentals, look in the local or English-language newspaper, check bulletin board postings at universities and local cafés, or enlist the help of a real estate agent. Make sure you have enough money to cover a deposit and several months' rent.

GETTING AROUND

Once you've safely arrived, slept a bit, and showered, you'll be anxious to start your adventure. You've waited for this moment. Now it's time to start exploring your new neighborhood, see some sights, and locate the good bakeries and pubs. But before you walk out the door, don't forget your handy guidebook and map. Write your temporary address inside, just in case you get lost and need to ask for help getting back. Believe me, it can happen, especially on your first day.

So, you walk out your door and head toward the old town square. On your map it seems pretty far away so you decide to take public transportation. Now you have to make sense of the local buses, trains, subways, and possibly even ferry systems. It can be daunting at first. But figuring out this confusing maze of transport will help you get oriented, and after a few days, you'll know the quickest way to get across town using a combination of transportation options. It is also fun to ask the help of locals, even if it means fumbling around in a foreign language. Most of the time people are more than happy to point you in the right direction. If you would rather be left to your own devices, your guidebook should give you some tips on where and how to buy transportation tickets.

The main train station is a good place to pick up a city map with train, bus, and other transportation routes. You'll have to learn to make sense of the colorful lines on the metro maps and get familiar with the strange bus-stop names. An information desk with an English-speaking attendant can also point you in the right direction. Once you know your daily route from home to school or work, you'll probably be able to buy a discount travel card that covers all modes of transportation you are using. Students, be sure to ask about student discount rates.

SETTING UP A BANK ACCOUNT

You'll need to set up a local bank account to sign up for utility services or to buy a cell phone. Always try to set up an account at an international bank that you know and trust, like Citibank. To be on the safe side, bring a letter from the head of your bank verifying the length of time you have had an account and that you are in good standing. Also, bring a few past bank statements to verify your banking activities. Depending on the country you are in, these items could expedite your getting a bank account established. You will also have to learn about the different banking practices and fees. When I lived in Germany, I never had to write checks for rent and utilities.

Instead, German banks have a system of paying bills automatically on a designated day. In Poland, by contrast, the banking system was not yet well established and most service payments were made in cash.

VISITING YOUR EMBASSY

It is a good idea to register with your embassy for your own safety and in case you need to be located in an emergency or for an evacuation. While you are there, check out postings for groups and clubs you can join. The American Embassy in Warsaw organized the Hash House Harriers, a running group known worldwide that is made up of people from various nations. And women can find out about professional women's clubs, which are a great place to network for jobs. Get involved as soon as possible.

BUYING A MOBILE PHONE

More people use cell phones in Europe than in the States, and service offerings are more advanced. Analyze the different phones and service plans to determine your best option. In Portugal, I use a prepaid phone service. I have a certain amount of phone usage time loaded on my mobile phone. As I make phone calls, the amount decreases and I can see on my display how much money I have left. Only outgoing calls are charged, not incoming. When I run out of money, I simply go to an ATM, and punch in my account number and the amount of money I want to load on the phone. Within a few minutes my phone is reloaded with a prepaid amount. I recommend this type of prepaid service if you want to keep track of how much you spend on phone calls. Portugal is one of the most advanced countries when it comes to prepaid services, so you may not find this option everywhere.

GETTING INTO THE RHYTHM

Whether you're living in Europe, Central America, or Asia, you'll eventually have to adjust to the local rhythms of life. Until you do that, your adventures might look a little like this. It's your first day and you're starving. You can't wait to try some local specialties. You go to a restaurant and there is something indecipherable scribbled on a note on the door. The next restaurant is open, but not a soul is in sight. The third restaurant has a few people sitting outside drinking beer. You start wondering where in the heck you can get some food. If it is 3:30 P.M., in many countries in Europe you might have to make do with downing a few beers until the restaurants open for dinner, around 6:00 P.M. in northern Europe and 8:00 P.M. in southern Europe. You'll learn the routine soon enough.

The next day you'll probably arrive at the restaurant around noontime. All the tables will be full and you might have to once again down a few beers waiting for a table to free up. Welcome to the European lunch hour, or hours, as the case may be. By the time you get some food in your stomach and you're ready to do some shopping, you discover that all the shops are closed during the lunch hour, and don't open again until around 2:00 or 3:00 P.M. You can try to do some other errands, like go to the bank or visit the embassy, but be sure to check their hours of operation because they might be closed for an extended lunch hour too.

On the third day, you get an early start to accomplish all the things you couldn't do in the first two days. You go to your train or subway stop and you wait, and wait, and check the schedule, and wait. It seems oddly empty. So you ask a passerby for help. He informs you it is one of the many public holidays. No one is working and the train and bus schedules have special hours, which you didn't notice on the schedule because you don't read the language yet. And since it is 8:30 in the morning on a national holiday, you again might be hard-pressed to find a café that is open that early.

After your disastrous morning, you go back home and call the one person you know in the city to clarify when you can actually eat, shop, and work in this country. Your friend is no doubt amused by your adventures, then he asks you to join him and his friends to do what most people do during the holidays. In Portugal that means heading straight to the beaches. In Denmark it can mean taking a cozy walk in the woods and having dinner with a small group of friends. In Germany, a free day off from work could be spent playing a game of soccer, followed by a few beers at a beer garden. And in Poland it could mean a drive to the seaside in the north, the lake region in the middle of the country, or the mountains in the south. As you can see, getting into the rhythm of the local culture takes time—anywhere from a few days to a few months. But once you know the local rhythm you'll know what to expect and life will get easier.

Maximizing Your Time Abroad

Many people want work overseas or be involved with global business, but equally important is life outside the office: practicing the language with natives after work at the local bar, backpacking around the region on weekends and holidays, socializing with an international circle of friends, getting involved with local issues, and participating in cultural events. Preparing for this kind of life is both fun and challenging.

Although it might seem unbelievable to adventure lovers like us, I have seen people miss out on the cool things that living in a foreign country offers, from exciting weekend trips and the hippest restaurants to traditional festivals and national events. Make sure you don't miss out on your experience of a lifetime—consider the suggestions below.

LEARN THE LANGUAGE

I've said it before, I'll say it again, learn the language! As much effort as it takes initially, it will be worth it to have the cultural insight and skills to get around later. Even if all you can do is order your favorite local wine or ask for directions, you'll be happy you could do it in their language.

READ WHAT THE LOCALS READ

In addition to investing in a good travel guide, you should read what the locals read, from the daily sports newspaper to the latest best-seller by a local author. This is the best way to gain additional insight into the culture. And, if you are up to date on local happenings, you'll have something in common with the people in your community, making it easier to strike up conversations with your neighbors, local shopkeepers, or the bartender at the corner pub.

GET INVOLVED WITH NATIVES AND OTHER FOREIGNERS

Getting to know the native population is often easier said than done. Take classes, join a health or sports club, and mingle with your work colleagues. And don't rule out other foreigners living abroad. They are there for similar reasons as you and sometimes can be more open to new friendships than the average local going about their daily business. Expat organizations can be located through Expat Exchange (www.expatexchange.com) or iAgora (www.iAgora.com). And, if you happen to be living in the U.K., get in touch with Focus Information Services (www.focus-info.org), which is run by expats and provides you with cross-cultural information and resources to make the most of your time there.

SIGHTSEE

Take advantage of the wonderful opportunity to travel locally. Try regional foods and attend local activities and festivals. Because you live in the country, you can do tours that travel companies would charge a fortune to do. Whenever guests visit us in Portugal, we take them on typical tourist routes through Lisbon and off the beaten path to lesser known medieval villages. Our itinerary would cost a few thousand dollars if you tried to arrange it through a travel agency, but we can do it for a few hundred dollars because we live locally.

Also take advantage of yearly festivals and events like Oktoberfest in Germany or Carnival in Brazil to get a taste of the national traditions. And, of course, visit the typical tourist sights, museums, markets, theaters, and monuments. They will give you plenty of insight into the history, art, and culture of that country.

GO ON HOLIDAY

Use your vacation time to sightsee outside of your home country. If you live in Europe, you'll quickly discover that European holidays are lengthier than American vacations. In Germany, for example, six weeks of vacation is the norm by law, even for entry-level employees. Many Germans travel outside of Germany and are known to be a "worldly," or well-traveled, culture because they can be found vacationing in all parts of the world.

In Poland, Portugal, and other countries of Eastern, Central, and Southern Europe, vacation days are typically about twenty-two to twenty-four days, which is also generous by American standards. Holiday time in other parts of the world can vary. In contrast to Germany the Japanese are known to have the least amount of vacation time—sometimes only a week or two. South Americans, Australians, and New Zealanders have about the same amount of holiday time as the average European—about four weeks.

RECORD YOUR MEMORIES

If you live abroad for the long term, you may forget to document your experiences because you stop feeling like a tourist. But, your time abroad could be an adventure of a lifetime, and you don't want to forget the wonderful people, places, and things you experienced. In addition to recording your memories on film and video, you can also remember your adventures in more unique ways. If you're artistic, make a drawing, sketch, or painting of your favorite scene—a beautiful cathedral, an ancient monument, a bustling marketplace. Or record your impressions of the culture and daily life in a diary. At the end of your travel adventure, make a scrapbook or collage with your favorite photos, keepsakes, and memorable moments.

STAY IN TOUCH

Write home frequently. Time doesn't stand still just because you're not there, and life will leave you behind, or at least out of the picture, if you don't stay in touch. Keep up with family, friends, and business colleagues at home to help combat feelings of homesickness and to hasten the readjustment process when you come back from your trip.

SAVE A FEW KEEPSAKES

Collect money, stamps, theme photos, brochures, or any type of souvenir that will remind you of your trip. I collected money from the different European countries I've visited because the Euro replaced many local currencies. Not only will the money eventually represent times past, it's a sentimental reminder of my various travel adventures.

KEEP A JOURNAL

Even if you don't write in a journal now, try starting one while you're abroad. You will never regret recording your experiences, especially if this is your first time overseas. Keeping a running account of the events that influence you during a particular time in your life can be very interesting to read in years to come. I like to record funny, sad, or historic stories, strange or exotic food recipes, my craziest moments, my most adventurous trip, local jokes and humor, my first impressions of people or countries, differences in mentalities, and even the local exchange rates. It is also fun to record the prices of things like a newspaper, a carton of milk, a telephone call, or a train ticket since costs will likely change drastically with time.

BRING GIFTS

A small gift from your homeland is a great way to start building lifelong relationships. If you are staying with a host family, bring a picture book of your own country to show them where you are from, or some local food and drink. And depending on where your travels take you, barter goods might come in handy. When my study-abroad class from Copenhagen visited Eastern Europe and Russia after the fall of the Berlin Wall, we brought cigarettes to pay for taxi cabs, candy for the kids, and items we knew the locals would want to trade for, like Levi jeans and perfume. In return, we received handpainted babushka dolls, Russian soldier uniforms, big furry hats, and worthless rubles. Having things to trade made it easy to strike up conversations with locals. We all learned a bit about each other and came away with treasured keepsakes.

The only word of caution I'll offer when it comes to gifts is to be aware of the effects your actions are having on the local culture. When my husband and I were trekking through Nepal, our local guides advised us *not* to give anything to the Nepalese children. Too many "generous" foreigners in the past had inadvertently bred a generation of children who learned that begging was a way of life. Instead, it was advised to simply smile, talk to them, or perhaps give them a Polaroid photo of themselves, as this was something unique.

GO WITH THE FLOW

Keep your expectations realistic and go with the flow. Adapt and make the best out of every situation. Let your experiences unfold before you instead of trying to control them. Stay healthy, exercise, laugh, have fun, and live.

Living overseas is a priceless education and experience if you make the most of it. Resist the urge to squeeze all the sight-seeing and shopping you didn't have time for into the last few stressful weeks of your stay. Enjoy each moment for all it's worth.

CHAPTER TWELVE

Home Sweet Home: Repatriation Reality

We must not cease from exploration.
And the end of all our exploring
will be to arrive where we began
and to know the place for the first time.

T. S. Eliot

When it comes time to start planning your trip home, don't be surprised if you feel lots of conflicting emotions. You're excited to see your old friends and family, but will miss the people who've become a big part of your life overseas. You can't wait to sink your teeth into your mom's pot roast, but can't imagine surviving without the savory delicacies you've grown to crave while abroad.

Upon returning home you will have to go through a period of readjustment to get back into the swing of things. You might find that you sync up with life at home quicker than you thought. Or you might experience a strange mixture of feelings that makes repatriation a slow and emotional challenge.

Coming Home

The best way to approach the process of leaving the new to return to the old is to be aware so you can prepare for the inevitable.

MAKE A PLAN FOR YOUR NEXT CHAPTER IN LIFE

If you are like most people, you'll engage in a bit of self-reflection when you realize you have to start thinking about going home. What were your original reasons for going overseas? Did you achieve what you set out to achieve? Are you satisfied with what you accomplished? What did you learn? Where do you want to go from here?

Whether you are finishing up a work assignment, study-abroad program, or long-term volunteer stint, now is the time to bring this exciting chapter in your life to a close. If you think of your life as a good novel filled with intriguing journeys and colorful encounters with people from around the world, then your story doesn't have to end just because you are changing your current surroundings. Start writing the next chapter from somewhere else in the world, even if that somewhere else is your home country. If you made a list of all the things you wanted to see and do while living abroad, now make a list of all the things you want to do when you return home, and spend as much time and energy trying to check the items off this new list as you did your original list. After all, going home can be an adventure too.

TAKE A REPATRIATION SEMINAR

Living abroad for a long period of time will leave its mark on you. You think differently now and you will inevitably experience a few bumps in the road to regaining balance in your life. Most expats initially feel a bit out of it when they first arrive home. It is not an uncommon phenomenon and help is available. Cross-cultural training companies like Windham International (www.windhamint.com) and Berlitz (www.berlitz.com) arrange repatriation seminars, and the Expat Exchange (www.expatexchange.com) has some good articles about repatriation on their Web site.

BUY SOUVENIRS AND GIFTS

If there is a favorite part about leaving for home, this is mine: a shopping rampage! Preparing to come home means a buying spree is in order before I begin packing. Stock up on items you won't be able to get in your part of the world. Buy all those things you meant to buy earlier. Take a trip to the outdoor market for bargains on clothes and handmade items. Stop by the bookstore to pick up picture books of the country as presents for friends at home. And finally, take one last trip to your local supermarket for your favorite goodies. And on the way home have an eating and drinking fest at all of your favorite restaurants and bars. Collect business cards from your favorite local joints to refer friends there in the future, or in case you return.

PACK UP

I don't know anyone who actually enjoys packing to go home. Not only must you pack up all the stuff you schlepped over, now you have to also find room for the new treasures you acquired. Remember to compare the cost of taking your extra things on the plane with sending them by a mail or moving company. If you decide to use a moving company, choose one that is a member of the internationally recognized FIDI moving association (www.fidi.com). In the four times we have moved overseas, the only time my precious pottery arrived intact was when we used a mover (Graebel Movers) approved by FIDI.

SAY GOOD-BYE

Organize a good-bye party to formally bid farewell to everyone who was part of your overseas experience. Be sure to write down their names and addresses so you won't lose touch with them. Although you might think you are leaving the country for good, you never know what will happen in the future. Staying in contact with the locals and other foreigners allows you to keep the doors of opportunity open.

I have said my final good-byes to Portugal three times in the last six years, each time thoroughly convinced that it would be a long time before I returned. But each time I ended up returning only a few months later.

RIDE THE WAVE

At some point during your move, you might feel a frustrating loss of control. Your belongings are in the hands of a foreign moving company, you no longer have a home here or there, and your social life has been virtually uprooted. You are floating free without anything to ground you. This is a perfectly normal symptom of repatriation.

And the only thing you can do about it is ride the wave. If you made a plan for how you will reanchor yourself once you set foot in home territory, you will soon regain a sense of control.

Repatriation Reality

When the time comes to return home you will undoubtedly have certain expectations about living in your home country again. Be aware that the reality you'll be facing at home could be different from the one you left, especially if you've lived overseas longer than a year.

BRAGGING OR SHARING?

When you get home you'll be bursting at the seams to share your adventures with your friends and family. However, in your enthusiasm to tell your stories and reconnect with friends, you might be perceived as a braggart or show-off. Although nothing could be further from the truth, it's up to you to be sensitive to this. People generally take a friendly, albeit fleeting, interest in your experiences, but sometimes they don't know what to say. Many travelers are disappointed when it seems like no one cares about where they've been living, what they've been working on, or all of the fun things they've done. People will ask you how your trip was, but few will have the patience to hear a detailed account of your adventures and look through picture albums full of people and places they don't know. Expecting your relatives and long-time friends to relate to your stories of far-off lands like Finland, Tajikistan, or the island of Yap could be asking too much, especially if they've never traveled farther than Mexico or Canada or Iowa.

Your friends' and family's disappointing response may leave you feeling a bit empty and isolated. In addition, you might feel pressured to move on with life at home before having time to digest and reflect on your overseas experience. Without time to share and reflect, it might seem like this exciting chunk of your life should be packed away and pushed to a tiny corner of your mind to be referred back to at some later point in time. And if this unexpected bursting of your bubble isn't enough to depress you, reverse culture shock (see page 55) might start to set in, especially if you have been abroad for several years or if it was your first time overseas. The key to a smooth assimilation is to get involved with global citizens who have their own memorable experiences to share. They'll understand exactly how you're feeling and will be excited to listen to your stories.

READJUSTING AT WORK

If you are a business professional who was sent abroad by your company, there's a good chance you'll have some adjustment issues when you go back to the home office. Some companies do not even guarantee you a position when your expatriate assignment ends. But even if you do have a job to return to, 80 percent of returned employees leave their companies within a few years, mainly due to the company's lackluster effort to repatriate the professional. Staying in touch with key colleagues in your home office, who will communicate your accomplishments to those who need to know, should help you maintain visibility and value in your company.

Another common issue for returning professionals is the realization that your work environment at home is less challenging. International assignments often include more responsibility and less direct control over your daily work flow. Unfortunately, things may be much different in the States. I know someone who was responsible for setting up a national office in Asia, including hiring staff and overseeing a multimillion-dollar budget. When he returned to his home office, he was stunned to find out he needed permission to hire a new secretary. The responsibility and control he had as a country manager overseas was simply dismissed in his new position at home. Ouch!

Indeed, it may be difficult to find a fulfilling role that will measure up to the constant challenge of the position you mastered overseas. And expatriates are generally able to save money, have more time off, and travel quite frequently. Furthermore, if your company paid for your house, car, and other expenses in addition to salary, stock options, and bonuses, it might be difficult to find a job that offers as much as your total expat package. If you want an expat-type package, working overseas might be the only way to get it. Otherwise, you'll have to mold your expectations to the current job market at home. That's reality.

A final challenge is when you find yourself job hunting in a changed business climate. This is precisely why keeping your network alive throughout your life is important. You never know when and who you may need to call on. You might have left during an economic boom and returned during a recession, or vice versa. Your company could have been restructured and key colleagues might have left the firm, or your company might have grown and added a lot of new employees with whom you must now build new relationships. Whatever challenge you face—finding a new job, building new relationships, leveraging your experience—your network is there to help. Just be sure you are able to return the favor.

Easing Your Transition

Before you find yourself up against a wall of frustration, realize that there are things you can do to ease your transition. Figuring out how to transform your global education into useful skills at home only requires a bit of creativity. Here are a few suggestions.

GIVE YOURSELF CREDIT

First of all, congratulate yourself. The personal characteristics it takes to succeed abroad combined with your unique experience of traveling and living overseas make you special. It takes a lot of effort to build a portfolio of international skills. Although international achievements are not yet consistently acknowledged, understood, and valued in American society, you've still accomplished something amazing that you should be proud of. You might have to give yourself a pat on the back if no one else will.

RECONNECT WITH FRIENDS AND BUSINESS CONTACTS

As soon as you know you're homeward bound, let your friends and business contacts know. If you are a social coordinator by nature, plan events to bring together your colleagues, friends, and others with global interests. Organize a dinner at a new ethnic restaurant. Plan your own coming-home party. Reestablish old ties. Your friends who have also traveled or lived abroad will relate to your latest travel adventures, making it easier for you to reconnect.

STAY INVOLVED WITH THE INTERNATIONAL COMMUNITY

Find an outlet for your new global skills. Surround yourself with like-minded people who seek personal fulfillment through involvement in international issues. Tapping into your local network of people with international experience will help you expand your contacts for future business and career opportunities. Your local chamber of commerce should be able to put you in touch with international businesspeople in your area. Local groups or clubs with international interests might be listed on your city's Web site or in the yellow pages. If you have children, find out if any international play groups are organized through local health centers or hospitals.

SEEK OUT CULTURE

Go to foreign films, ethnic restaurants, museums, or events where you can continue to expand your level of cultural appreciation. Meet with friends who have traveled or foreigners in your town to exchange travel stories, share pictures, and show videos.

MAINTAIN YOUR LANGUAGE SKILLS

If you ever plan on using your foreign language skills again, you must maintain them. Take advanced or conversational courses at a local community college or university. Or seek out community groups that get together on a regular basis to practice language. Language departments can sometimes recommend these groups. I have been able to find groups of Germans and foreigners who speak German that meet regularly in a monthly *Stammtisch* to drink beer and speak German. It's always a good time and an easy way to make friends.

USE YOUR NEW SKILLS

Use your experience abroad to start a part-time business or a new career. Become a travel consultant, give presentations about the importance of an international education, write articles about the country you lived in, develop an online business to connect global business contacts with each other, or mentor professionals who aspire to walk in your global shoes. Your local network of international friends might have other suggestions for you too.

SET UP BUSINESSES OVERSEAS

There are several volunteer organizations that send seasoned businesspeople abroad to help developing countries set up businesses and negotiate international contracts, and to act as business liaisons. If you've got these business skills, volunteer your expertise on such projects. For more information, check out the International Executive Service Corps (see page 189) or the Citizens Democracy Corps (see page 185). Your local chamber of commerce may also have information about businesses that are developing globally.

BECOME A MENTOR

If you studied or interned overseas, offer to give a talk about your experiences at your school or mentor other students who want to study and work abroad. If you're no longer a student, talk to high school guidance counselors or college advisors about working with students. If you're a business professional, offer to mentor colleagues who want to develop global careers. Your human resource and career development departments can probably help you develop a program.

JOIN A PROFESSIONAL ORGANIZATION

To maintain your professional skills and contacts, join a professional organization. Most professional organizations have international branches and sponsor events that bring together people who are involved in your profession from all over the world.

SPEAK AT A CHAMBER OF COMMERCE EVENT

Offer your expertise to the local chamber of commerce. Speak at one of their events or organize a seminar of your own. At the very least, maintain your chamber of commerce contacts, as they are important for keeping you in the networking loop.

Repatriation does not have to be difficult. Returning home after months or years of living an international life can simply be a continuation of the global life you started overseas. Consider your trip home the next step toward becoming part of the international community.

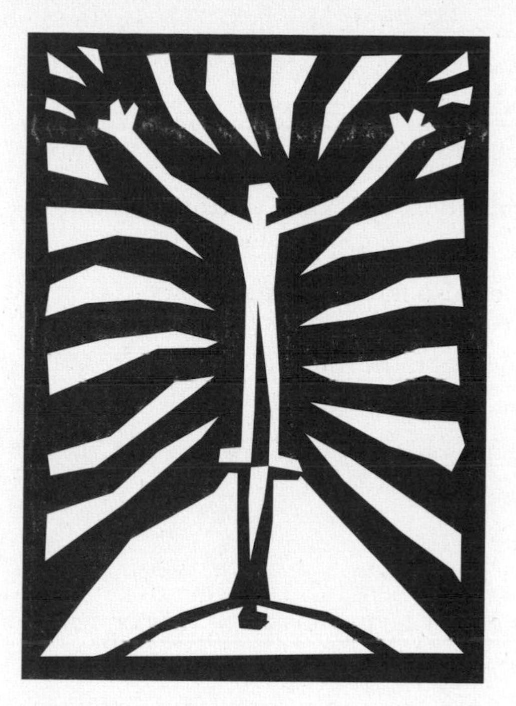

CHAPTER THIRTEEN

Contributing to the World as a Global Citizen

We make a living by what we get,
but we make a life by what we give.
NORMAN MACEWAN

Whether you are at the beginning of your global journey or a seasoned traveler adept at adjusting to life in different countries, at some point you might wonder how you can pass on what you've learned, influence others to make the world a better place, and contribute as a citizen of the world.

Living abroad for an extended period of time makes you aware of the issues facing the country and region in which you are living. By the same token, it puts you and your home country's issues into perspective on the world scene. As you learn to ignore stereotypes and become less ethnocentric in your worldview, your concern for other citizens, the environment, and global causes may broaden. International involvement is needed to help humankind increase cultural and environmental awareness and tolerance, and solve economic, political, and human rights problems.

Accepting and understanding cultures and other ethnic groups is a step in the right direction for world peace and an end to war, hunger, poverty, crime, and environmental destruction. Many of the educational travel, study, work, and volunteer organizations listed in this book offer great opportunities to get involved as you take responsibility for being a citizen of the world.

What Can You Do?

At some point during your journey, spurred by the insights you gain from your global experience, you may realize that you feel most fulfilled when you're making a difference for someone else. It is a privilege to be able to travel and experience other cultures and the world at large, and many people find it gratifying to give back to the societies that have taught them what they now know.

JOIN

Become a member of an organization or subscribe to journals or magazines that use your donations to serve a cause. This can be as simple as subscribing to *National Geographic* or as involved as becoming an active member of Greenpeace. If you have friends or relatives who are interested in the environment or wildlife, a gift membership is a great way to make a valuable contribution while helping them develop their interests. Subscribe to member newsletters to stay up to date on the latest developments with your favorite causes. Many of the volunteer and exchange organizations listed in this book publish newsletters on a variety of topics relevant to their missions.

DONATE

Donating money and volunteering your time are more traditional ways to contribute to your favorite cause. Many organizations will gratefully accept a gift of stock, which entitles you to tax savings through a charitable deduction, or regular monthly contributions and bequests. Convince your company to join in the effort through corporate gift matching, contributions, or the organization of volunteer activities. You can also donate untraditional items such as frequent-flier miles and mobile telephones. Both individual and corporate contributions are tax-deductible.

Becoming an activist for your favorite cause means you actively help the organization reach its goals. Join an expedition to raise awareness for CARE or participate in an Amnesty International letter-writing campaign to stop torture. Your company might also be interested in getting involved with events to raise money for special causes.

The following list is only a small selection of the thousands of organizations you can join to contribute to worldwide causes.

★ **AMNESTY INTERNATIONAL/AMNESTY USA**

www.amnesty.org or www.aiusa.org

Amnesty International USA
322 Eighth Avenue
New York, NY 10001
Phone: 212-807-8400
Fax: 212-627-1451

Join: Amnesty International is a Nobel Prize–winning grassroots activist organization with over one million members worldwide. Amnesty works to free prisoners of conscience, assure fair trials for political prisoners, and bring an end to torture, political killing, "disappearances," and the death penalty. If you are a U.S. citizen, Amnesty memberships cost a minimum of $25 USD ($15 USD for students, seniors, and people with limited incomes). Members receive an official Amnesty International membership card, a free subscription to their quarterly news magazine *Amnesty Now,* and the right to vote in the annual election of Amnesty International's board of directors. Membership contributions allow Amnesty to work without the financial assistance of governments, thus allowing them to operate as an impartial and outspoken advocate.

Donate: In addition to annual membership dues, one-time contributions for special occasions or to honor someone close to you are accepted. Gifts of stock, bequests, and planned giving are other ways to donate. Or sign up for Amnesty's checks, long-distance phone service, or credit card.

Act: Amnesty International activists make a difference by pressuring human rights violators worldwide. Actions you can participate in are listed on the Web site. Join a local group to "adopt" a prisoner of conscience, email your representative, or write appeals for worldwide victims of human rights violations such as rape and torture. See Amnesty's Stop Torture Campaign at www.stoptorture.org. Get involved and tell others about it.

★ CARE International

www.care.org/support/gift

CARE USA
151 Ellis Street, NE
Atlanta, GA 30303
Phone: 404-681-2552
Fax: 404-577-5977

Join: Although CARE doesn't offer traditional memberships, there are numerous ways to donate your time and money.

Donate: CARE is one of the world's largest private international relief and development organizations, with ten agencies operating in fifty countries. You can contribute to CARE's work in many ways whether you are an individual or you represent an organization. Make monetary donations online, by phone, or through the mail. Check with your company's personnel office to see if they participate in a matching-gifts program. Many companies will match, double, or sometimes triple tax-deductible contributions made by employees. If you've got frequent-flier miles or a used mobile phone, why not donate them? And, if you are in the position to do so, give cash, securities, or real estate, all of which offer you tax benefits.

Act: Climb for CARE is a program to raise money and increase awareness for CARE's humanitarian work worldwide by climbing Mount Kilimanjaro in Tanzania. Information can be found on the CARE Web site. There are also special projects to support women's rights. Or you can support CARE's relief and development efforts by shopping at CARE's marketplace, where you can buy books and items made with CARE products. Three to 12 percent of the purchase price is donated to CARE. To see how CARE operates in your community, check their Web site for more information.

★ Earthwatch Institute

www.earthwatch.org

680 Mt. Auburn Street, Box 9104
Watertown, MA 02471
Toll-free: 800-776-0188
Phone: 617-926-8200
Fax: 617-926-8532

Join: Earthwatch's mission is to promote "sustainable conservation of our natural resources and cultural heritage by creating partnerships between scientists, educators, and the general public." Earthwatch members support more than 140 scientists on research expeditions in fifty-five countries. There are different levels of membership and costs vary per country. An annual individual U.S. membership is $35 USD and includes an *EWI* journal subscription, a research and exploration guide, expedition and event discounts, access to the Earthwatch Institute network, an online newsletter, and an L.L.Bean traveler gift certificate.

Donate: You can make a cash donation to Earthwatch in support of cultural diversity research and other programs, or make an indirect contribution by buying Earthwatch merchandise from its partner organizations like L.L.Bean. A portion of your purchase is used to fund Earthwatch projects.

Act: Join an Earthwatch project (see page 199). Volunteers are actively involved in conservation projects, working hand in hand with indigenous peoples to understand their problems and find solutions.

★ Global Exchange

www.globalexchange.org

2017 Mission Street, #303
San Francisco, CA 94110
Toll-free: 800-497-1994
Phone: 415-255-7296
Fax: 415-255-7296

Join: When you become a member of Global Exchange, you support an international grassroots movement that builds people-to-people ties. Members receive a quarterly newsletter and action alerts, priority consideration for Reality Tours (see page 109), discounts at Fair Trade Crafts Centers, and connections to a large network of activists. Memberships cost $35, $50, and $100 USD, depending on your level of membership.

Donate: Global Exchange accepts donations in any amount. If you would like to donate your time rather than money, Global Exchange needs volunteers and interns to work in their San Francisco office or on special projects supporting human rights, democracy, and other issues. Global Exchange also sells products from the destinations in which they work. A portion of the sale is donated directly back to the community.

Act: On the Action Center page of their Web site you can choose issues on which you'd like to act by writing a letter, signing a petition, or making a call to show your support. Global Exchange maintains several email lists on various topics. For example, "Colombia News" provides news and action alerts on human rights and U.S. military aid in Colombia. Similar email lists exist for Cuba, Mexico, and Palestine. As part of Global Exchange's mission to educate people about social justice worldwide, they also offer Reality Tours, which are speaking tours conducted by activists struggling for justice in their home countries, and publish pamphlets and other resources for the public.

★ **GREENPEACE**

www.greenpeace.org

702 H Street NW
Washington, DC 20001
Toll-free: 800-326-0959

Join: To help Greenpeace fight in the struggle for a green and peaceful planet, make a single donation of $25 USD or more and receive a one-year subscription to their international newsletter. If you sign up for the Rainbow Warrior monthly giving plan, a minimum of $10 USD per month entitles you to the international newsletter for a year, an annual report, and a Greenpeace video.

Donate: Donations of any amount are accepted. Payments made in U.S. dollars are preferred.

Act: Greenpeace is known worldwide for its activities to protect the environment. You can take part in the Greenpeace Cyberactivists online discussion, join a mailing list, or become a Greenpeace supporter.

★ Health Volunteers Overseas (HVO)

www.hvousa.org

P.O. Box 65157
Washington, DC 20035
Phone: 202-296-0928
Fax: 202-296-8018

Join: To support HVO's efforts to provide proper health and medical education around the world, make a tax-deductible contribution according to your professional status. The minimum suggested amount is $25 USD for health-care professionals in training and gradually increases from there. For an additional $15 USD, you can purchase their book, *A Guide to Volunteering Overseas,* or an HVO poster.

Donate: HVO gladly accepts your tax-deductible donations in any amount. Their Web site has an article clearly explaining how planned giving through a charitable remainder trust (CRT) is a smart tax-planning strategy.

Act: Volunteering with HVO (see page 202) is the most common way to participate. Qualified health-care professionals are sent abroad to train their counterparts in one of fifty projects in twenty countries throughout Africa, Asia, Latin America, Eastern Europe, and the Caribbean.

★ National Geographic Society

www.nationalgeographic.com

1145 17th Street, NW
Washington, DC 20036
Toll-free: 800-647-5463
Phone: 813-979-6845

Join: By subscribing to *National Geographic* magazine, you help the National Geographic Society explore, protect, and educate. A portion of the $29 USD subscription fee is used for international research and exploration. In addition to *National Geographic* magazine, there are a host of other books, magazines, CD-ROMs, and merchandise on topics as varied as Mount Everest and the world's largest flower.

Donate: In support of National Geographic's work in research, exploration, and geography education for young people, contributions of $25 USD or more will receive *Connections,* the society's newsletter for donors that includes a behind-the-scenes look at the researchers, explorers, and expeditions. Bequests and planned gifts are also accepted.

Act: To get involved with National Geographic's mission, you can travel with a National Geographic team on an expedition, which is funded in part by membership donations. National Geographic is also looking for affiliates who can promote their store products and spread the word about National Geographic's activities. Affiliates even receive a percentage of the sales. See the store affiliate program on the Web site for details.

★ Peace Brigades International (PBI)

www.igc.apc.org/pbi

428 8th Street, SE, 2nd Floor
Washington, DC 20003
Phone: 202-544-3765
Fax: 202-544-3766

Join: PBI is a grassroots, nonviolent peacekeeping organization that promotes human rights worldwide. They currently have projects in Colombia, Indonesia/East Timor, and Mexico. The suggested membership contribution is $30 USD.

Donate: PBI volunteers work long days in dangerous areas to keep local human rights activists safe. Your tax-deductible donations of any amount support volunteers in the field. You can choose to have your funds electronically transferred to PBI, thus reducing administrative costs; make a donation of appreciated stock, which will have tax advantages for you; or designate PBI as a beneficiary in your will.

Act: PBI volunteers and supporters courageously work together to promote peace in regions where governments are not willing to act. You can show your support by joining the Emergency Response Network, volunteering on a PBI project, participating in a delegation, helping with public relations and outreach, making a financial contribution, setting up a new project, or developing a PBI group in your country if one doesn't already exist.

★ Women for Women International

www.womenforwomen.org

733 15th Street, NW, Suite 310
Washington, DC 20005
Phone: 202-737-7705

Join: Although Women for Women doesn't offer traditional memberships, there are many ways to donate your time and money.

Donate: Women for Women is a nonprofit humanitarian organization devoted to the financial, educational, and interpersonal support of female survivors of war, poverty, and injustice. Women for Women provides resources women need to move out of crisis into positions of stability and self-sufficiency. Donations of any amount are gratefully accepted, however, they can only be made in U.S. dollars. You can choose how your cash contribution is spent by selecting a program fund or other category.

Act: Get involved by writing letters of support to women around the world, making a monthly donation, volunteering, or interning. Volunteer positions may be available in Bangladesh, Bosnia, Kosovo, Nigeria, and Rwanda. If you are fluent in Albanian, Bangla, Bosnian/Serbo-Croatian, French, or Kinyrwandan, Women for Women is in need of volunteers to translate. In addition, you can join Women for Women International's global family by sponsoring a woman in need, helping women renew their life skills to become self-sufficient, or donating to a microlending program or country fund. Furthermore, your purchase of handcrafted items generates income for Women for Women International. See the Web site for more information.

★ The World Wildlife Fund (WWF)

www.wwf.org

1250 24th Street, NW
Washington, DC 20037
Toll-free: 800-CALL-WWF

Join: WWF is the world's largest and most experienced independent conservation organization. Its mission is to stop the destruction of the natural world and protect the biological diversity that we need to survive. Membership donations range from $15 to $5,000 USD or more, and gift memberships for family and friends are also available. Membership includes a free T-shirt and a subscription to the *Living Planet* magazine.

Donate: Donations of any amount are welcome. You can also give gifts from your stocks, retirement plans, insurance policies, and savings accounts, or include WWF in your will. Corporations can make philanthropic donations of $1,000 USD or more. Or, if your company would like to support WWF through products, services, and promotions that raise funds to increase WWF's conservation efforts, you may be able to arrange a marketing partnership.

Act: Join WWF's Conservation Action Network and take action on issues related to wildlife and environmental conservation. Membership is free; you simply reply and add your comments to monthly "action alert" emails, which are then forwarded to the appropriate decision makers. You can also become a member of the Wildlife Rescue Team. Your regular monthly contributions make a significant impact on WWF's worldwide conservation efforts. You'll receive the bimonthly newsletter called *FOCUS* and regular updates from the field.

BUY GIFTS TO SUPPORT CAUSES

The next time you are looking for a unique present for someone, considering buying gift baskets, crafts, food, and books from organizations that filter a portion of the profits back to the communities where the products were made. Whether or not you are a member of the listed organizations, you can still buy products to support their causes and benefit the communities in which they work.

★ Global Exchange

www.globalexchange.org

On the Global Exchange Web site you can buy products purchased directly from the artisans. A Guatemalan baby sling is $60 USD, Kenyan carved candleholders are $11 USD, or a Salvadoran handpainted letter holder is $15 USD. Proceeds support the communities that produced the items.

★ Women for Women International

www.womenforwomen.org

A pair of Privlagci slippers, snugly handmade slippers that are an important symbol of Bosnian hospitality and warmth, cost $20 USD per pair, plus shipping.

★ The World Wildlife Fund

www.wwf.org

WWF's online store sells giant stuffed pandas, tigers, and elephants; binocular kits; umbrellas; backpacks; checks with wildlife themes; and more. A portion of the proceeds are used to support WWF's mission to protect wildlife and the environment.

SPONSOR AN EXCHANGE STUDENT

Sponsoring an exchange student is an excellent way to contribute to international exchange and understanding. Open your home to a student from a foreign country to give your children interaction with a foreign student and to learn about another country from a native. Exchange organizations like Council on International Educational Exchange (see page 127) can help you arrange the details.

SHARE YOUR KNOWLEDGE

As you discover more about yourself and the world, pass on your knowledge to others. Write articles for the local newspaper or a company newsletter; speak at the local high school, chamber of commerce, or other organization; or give cross-cultural training suggestions to the human resources department at work. Teach a class, act as a mentor to young people who want to gain international experience, or share your stories on a message board. Passing on your global knowledge and wisdom helps us all move toward a world of global understanding and peace.

Joining organizations that reflect your priorities in life, donating to your favorite causes, becoming an activist, sponsoring an exchange student, and sharing your experiences with others are just a few of the numerous ways you can make the world a better place. Trying to give as much back to the world as the world has given you is the mark of a true global citizen.

Bon Voyage

Every country I've lived in has taught me something new about myself and the world. Studying, working, and traveling abroad have opened up worlds of opportunity that I would never have been aware of if I hadn't taken risks. To accommodate my transient lifestyle I've become more flexible, entrepreneurial, and much less traditionally career-oriented than I ever anticipated. My priorities at this stage of life are to see as much of the world as possible, become fluent in a few languages, and build professional skills that will keep me marketable in the global workforce.

I have had the unique privilege to live and work in five different countries; become proficient in German, Portuguese, and Spanish; and travel to over thirty countries. I have been enlightened by what this world has to offer and I hope you will be too. A wonderful aspect about being on an international path is that every person's path is unique. Know yourself and make the most of your potential. While you were reading this book, I hope you have been inspired by what the world has to offer. Whether you have a one-time experience abroad or continue to pursue international opportunities for the rest of your life, your overseas adventures will set you apart by shaping your personal and professional development and leading to a fulfilling life rich with exciting opportunities.

It is time to begin your global journey . . . Bon voyage!

Tribute

Over the course of the last twelve years of my travels, and during the writing of this book, I have lost numerous close family members and loved ones. It is in their untimely deaths that I have learned to truly live and appreciate every moment of life. Their courageous struggles have empowered me to write this book, and their passion for living is reflected in the pages you've just read. This book is in loving memory of those relatives and friends, in particular my late cousins Julie Ann Iocco (1968–2000) and Jean Marie Iocco (1975–2001).

About the Author

As an American expatriate (and spouse of a German expatriate), Elizabeth Graham Kruempelmann has spent the last twelve years studying and working in the United States, Denmark, Germany, Poland, and Portugal. She has traveled to over thirty countries in Europe, Asia, Africa, and the Middle East, and speaks English, German, and Portuguese. Elizabeth currently works as a writer, cross-cultural sales consultant, and mother. Previously, she worked as an international message board host for Monster.com, an English instructor at Berlitz, and a partner in a small advertising agency in Poland, where she witnessed Poland's dramatic transition to a market economy. A native of Corning, New York, and a graduate of SUNY Albany, Elizabeth lives in Portugal with her husband and son, with baby #2 on the way.

Elizabeth welcomes mail from readers. Please feel free to email her directly at ekruempe@hotmail.com or contact her through *The Global Citizen* Web site at www.the-global-citizen.com.

Index